CAPACITANCE, INDUCTANCE, AND CROSSTALK ANALYSIS

The Artech House Microwave Library

Analysis, Design, and Applications of Fin Lines, Bharathi Bhat and Shiban K. Koul
E-Plane Integrated Circuits, P. Bhartia and P. Pramanick, eds.
Filters with Helical and Folded Helical Resonators, Peter Vizmuller
GaAs MESFET Circuit Design, Robert A. Soares, ed.
Gallium Arsenide Processing Techniques, Ralph Williams
Handbook of Microwave Integrated Circuits, Reinmut K. Hoffmann
Handbook for the Mechanical Tolerancing of Waveguide Components, W.B.W. Alison
Handbook of Microwave Testing, Thomas S. Laverghetta
High Power Microwave Sources, Victor Granatstien and Igor Alexeff, eds.
Introduction to Microwaves, Fred E. Gardiol
LOSLIN: Lossy Line Calculation Software and User's Manual, Fred. E. Gardiol
Lossy Transmission Lines, Fred E. Gardiol
Materials Handbook for Hybrid Microelectronics, J.A. King, ed.
Microstrip Antenna Design, K.C. Gupta and A. Benalla, eds.
Microstrip Lines and Slotlines, K.C. Gupta, R. Garg, and I.J. Bahl
Microwave Engineer's Handbook: 2 volume set, Theodore Saad, ed.
Microwave Filters, Impedance Matching Networks, and Coupling Structures, G.L. Matthaei, L. Young and E.M.T. Jones
Microwave Integrated Circuits, Jeffrey Frey and Kul Bhasin, eds.
Microwaves Made Simple: Principles and Applications, Stephen W. Cheung, Frederick H. Levien, et al.
Microwave and Millimeter Wave Heterostructure Transistors and Applications, F. Ali, ed.
Microwave Mixers, Stephen A. Maas
Microwave Transition Design, Jamal S. Izadian and Shahin M. Izadian
Microwave Transmission Line Filters, J.A.G. Malherbe
Microwave Transmission Line Couplers, J.A.G. Malherbe
Microwave Tubes, A.S. Gilmour, Jr.
MMIC Design: GaAs FETs and HEMTs, Peter H. Ladbrooke
Modern Spectrum Analyzer Theory and Applications, Morris Engelson
Monolithic Microwave Integrated Circuits: Technology and Design, Ravender Goyal, et al.
Nonlinear Microwave Circuits, Stephen A. Maas
Terrestrial Digital Microwave Communications, Ferdo Ivanek, et al.

CAPACITANCE, INDUCTANCE, AND CROSSTALK ANALYSIS

Charles S. Walker

Artech House
Boston • London

Library of Congress Cataloging-in-Publication Data

Walker, Charles S.
    Capacitance, inductance, and crosstalk analysis / Charles S. Walker.
    p. cm.
    Includes bibliographical references.
    ISBN 0-89006-392-3
    1. Electric circuits. 2. Electric inductors. 3. Capacitors. 4. Crosstalk. I. Title.
TK454.W36   1990   90-252
621.319'21--dc20   CIP

British Library Cataloguing in Publication Data

Walker, Charles S.
    Capacitance, inductance, and crosstalk analysis.
    1. Electromagnetism
    I. Title
    537

    ISBN 0-89006-392-3

© 1990 ARTECH HOUSE, Inc.

685 Canton Street
Norwood, MA 02062

All rights reserved. Printed and bound in the United States of America. No part of this publication may be reproduced or utilized in any form or by any means, electronic or mechanical, including photocopying, recording, or by any information storage and retrieval system, without permission in writing from the publisher.

International Standard Book Number: 0-89006-392-3
Library of Congress Catalog Card Number: 90-252

10 9 8 7 6 5 4 3 2 1

To
Catherine and Our Daughters

Contents

Preface	xi
Introduction	xiii

Chapter 1   General Concepts — 1
- 1.1   Introduction — 1
- 1.2   Resistors, Capacitors, and Inductors — 1
  - 1.2.1   Resistors — 2
  - 1.2.2   Capacitors — 6
  - 1.2.3   Inductors — 10
- 1.3   Electric Field Mapping — 15
  - 1.3.1   Parallel Plate Capacitor — 15
  - 1.3.2   Circular Conductor Capacitor — 17
  - 1.3.3   Elliptical-Rectangular Conductor Capacitor — 20
  - 1.3.4   Rectangular Conductor between Two Ground Planes — 20
- 1.4   The $LCRZ_0$ Analogy — 22

Chapter 2   Formula Sets — 31
- 2.1   Introduction — 31
- 2.2   Capacitance — 32
  - 2.2.1   Capacitance between Two Circular Conductors, C-1 — 32
  - 2.2.2   Capacitance between a Circular Conductor and a Ground Plane, C-2 — 36
  - 2.2.3   Two-Circular-Conductor Mutual Capacitance Near a Ground Plane, C-3 — 39
  - 2.2.4   Capacitance between Parallel, Vertical, Flat Conductors, C-4 — 48
  - 2.2.5   Capacitance between Horizontal Flat Conductors, C-5 — 51
  - 2.2.6   Capacitance between a Flat Conductor and a Ground Plane, C-6 — 55
  - 2.2.7   Mutual Capacitance between Two Horizontal, Flat Conductors Near a Ground Plane, C-7 — 62

	2.2.8	Four-Conductor-System Mutual Capacitance, C-8	66
	2.2.9	Capacitance between a Flat Conductor and Two Ground Planes (Stripline), C-9	71
	2.2.10	Mutual Capacitance between Two Flat Conductors Near Two Ground Planes, C-10	79
	2.2.11	Capacitance of Coaxial Cables, C-11	81
	2.2.12	Capacitance between Two Small Spheres, C-12	83
2.3	Inductance		85
	2.3.1	Self-Inductance of Two Circular Conductors, L-1	85
	2.3.2	Self-Inductance of a Circular Conductor and a Ground Plane, L-2	88
	2.3.3	Mutual Inductance between Two Conductors Near a Ground Plane, L-3	89
	2.3.4	Self-Inductance of Vertical, Flat Conductors, L-4	92
	2.3.5	Self-Inductance of Two Horizontal, Flat Conductors, L-5	95
	2.3.6	Self-Inductance of a Long Flat Conductor and a Ground Plane, L-6	97
	2.3.7	Mutual Inductance between Two Flat Conductors Near a Ground Plane, L-7	100
	2.3.8	Four-Conductor-System Mutual Inductance, L-8	101
	2.3.9	Self-Inductance of a Stripline (Flat Conductor between Two Ground Planes), L-9	110
	2.3.10	Mutual Inductance between Two Flat Conductors Near Two Ground Planes, L-10	113
	2.3.11	Self-Inductance of Coaxial Cables, L-11	115
	2.3.12	Self-Inductance of Circular and Square Loops, L-12	118
2.4	Characteristic Impedance for Various Geometries, Z_0-(ALL)		121
Chapter 3	Crosstalk Analysis		125
3.1	Introduction		125
3.2	Capacitive Crosstalk		125
	3.2.1	Capacitive Coupling to Summing Junctions, CTC-1	125
	3.2.2	Capacitive Coupling to Summing Junctions (with Ground Plane), CTC-2	133
3.3	Inductive Coupling between Circuits, CTL-1		135
3.4	Common Ground Coupling, CTG-1		140
3.5	Circuit Crosstalk Due to Power Supplies, CTPS-1		149
Chapter 4	Discrete Components		153
4.1	Introduction		153
4.2	Characteristics of Commonly Used Capacitors, DC-1		153
4.3	Characteristics of Commonly Used Inductors, DL-1		160
Chapter 5	Ancillary Circuit Elements		165
5.1	Introduction		165

5.2	Printed Wiring Board Resistance		165
	5.2.1 Conductor Resistance, R-1		165
	5.2.2 Ground Plane Resistance, R-2		167
5.3	Voltage Sources, V-1		171
5.4	Current Sources and Sinks, I-1		172
5.5	Power Supply Charateristics, PS-1		173

Chapter 6 Experiments and Test Data — 177

6.1	Introduction		177
	6.1.1 Experimental Results		177
	6.1.2 Circuit Board Hardware Design		177
	6.1.3 Circuit Board Schematics		178
	6.1.4 Test Equipment		178
6.2	Capacitive Crosstalk		178
	6.2.1 Capacitance between Parallel, Vertical, Flat Conductors, EXP C-4		178
	6.2.2 Capacitance between Parallel, Horizontal, Flat Conductors, EXP C-5A		188
	6.2.3 Horizontal Flat Conductors with Guard Rings, EXP C-5B		191
	6.2.4 Mutual Capacitance of Two Parallel, Horizontal, Flat Conductors Near a Flat Ground Plane, EXP C-7		198
6.3	Inductive Crosstalk		201
	6.3.1 Four-Conductor-System Mutual Inductance, EXP L-8		201
6.4	Ground Return Crosstalk		206
	6.4.1 Shared-Ground Crosstalk, EXP CTG-1A		206
	6.4.2 Single-Point Ground Crosstalk, EXP CTG-1B		208
	6.4.3 Downstream Power Supply Crosstalk, EXP CTG-1C		209

Appendix A	213
Bibliography	223
List of Symbols	225
Index	227
The Author	231

Preface

The purpose of this book is to fill several voids in the field of electromagnetics for both the practicing engineer and the electrical engineering student.

From direct experience and that of colleagues, electromagnetics is viewed as one of the most difficult to understand courses taught in the universities. The highly conceptual nature of the subject combined with the complications of three-dimensional vector calculus produces a topic that is not easy to master or retain.

Chapter 1 discusses two-dimensional field theory in simplified terms with figures showing parallel-plate resistors, capacitors, and inductors. Flux plots provide visual images of electric fields and are intended to give the reader a grasp of the basic mechanisms involved.

Equations and formulas giving the values for inductance, capacitance, and characteristic impedance are available in the literature, but are widely scattered. Chapter 2 presents formula approximations for common circuit geometries. With the exception of several very simple geometries, exact solutions for capacitance and inductance values are exceedingly complex. The thrust of Chapter 2 is to give the engineer easy access to equations that provide quick solutions with adequate accuracy for most electrical engineering problems. Both SI (International System of Units) and English units of measurement are used. Examples and measurements support the equations providing better insight for the size of the quantities involved.

The adverse effects of crosstalk and noise present one of the most frustrating and difficult-to-solve problems in electrical engineering. Chapter 3 describes methods for attacking crosstalk problems. It discusses the various ways crosstalk can be caused and provides quantitative design equations for predicting crosstalk levels.

Capacitors and inductors, as discrete components, play important roles in electrical engineering. Chapter 4 is devoted to an analysis of these components and their limitations. Measured data are presented to validate and quantify the equivalent circuits.

Chapter 5 presents ancillary circuit elements and concepts. These include the evaluation of circuit board land and ground plane resistances, voltage-current sources, and power supply characteristics.

Test data provide essential feedback in electronic circuit design. Most circuits are "breadboarded" prior to final design for production. This breadboard provides the engineer the assurances needed to complete the circuit designs. Chapter 6 discusses eleven experiments specifically designed to demonstrate the concepts and verify the equations presented in Chapters 2 and 3. Predicted values are calculated and then compared with measured data.

For convenience, a List of Symbols is presented at the end of this book.

The author has taken reasonable steps to ensure the accuracy of the formulas, tables, and other material in this book, as illustrated by the large number of worked-out examples supported by empirical test data and independent references. However, due to the complex nature of this subject area, he cannot assume the responsibility that the material is totally error-free or subject to misinterpretation. If errors are found, please let us know and we will correct them in subsequent printings.

This book, in part, is an extension of studies conducted to help explain and prevent the severe crosstalk problems we had experienced on printed circuit boards. Engineering notes were formalized as a set of Design Guidelines. Selected laboratory experiments, conducted later, verified study findings. I want to thank and recognize Honeywell, Inc., Marine Systems Division, for their support in this effort. Thanks are also extended to Marshall Okamoto and Tom Ramus for conducting the experiments; to Carol Leupold, who spent many hours typing the lion's share of the original design guidelines.

Special thanks are given to Professor John L. Bjorkstam, University of Washington, who has offered very helpful suggestions for the theoretical portions of this book, and to Ken Sagara, who prepared many of the illustrations.

This effort has involved my entire family and I wish to express my appreciation for their contributions. Daughters Ann and Mary furnished much of the draft typing and other support for the manuscript. Daughters Elizabeth, Megan, Stephanie, and Ellen took care of many tasks, normally my responsibility, during the preparation of this book. My nephew, Roderic O'Quigley, helped proofread the galleys. Last, but not least, thanks go to my wife, Catherine, for her continuing encouragement and support of this project.

CHARLES S. WALKER
SEATTLE, WASHINGTON
JANUARY 1990

Introduction

This book provides electrical engineers and students with basic electromagnetic theory and formulas for application to engineering problems. It is not intended to replace standard college text books on electromagnetic theory—rather it is designed to give a basic understanding of the concepts involved, and provide formulas for the solutions of crosstalk problems on circuit boards and other areas of electrical engineering. Formulas and analysis techniques are presented in practical and easy-to-use terms with examples comparing relative merits of different approaches.

Exact field problem solutions for all but the most elementary geometries are exceedingly difficult, time consuming and, in some cases, impossible. In this book, we build on simple geometries to establish values. This method yields approximate solutions with accuracies sufficient for most engineering problems. Oftentimes, we find that circuit configurations cannot possibly meet the performance requirements. For example, the maximum crosstalk requirement between channels might be specified to not exceed -30 dB. A simple calculation would show that this value could not be achieved unless a ground plane were used.

This book starts with relatively simple circuit geometries and explores the effect of these geometries on circuit electrical performance. More difficult geometries are then investigated. Before starting the design of printed wiring board or ceramic module layouts, the engineer can first refer to the chapter on crosstalk analysis and review various crosstalk causes. The formula sets are then consulted to determine expected circuit parameter values such as mutual capacitance or inductance. Calculations are then made to determine the adequacy of the proposed design. Review of Chapter 6, Experiments and Test Data, provides concept reinforcement.

Each formula set and crosstalk analysis is based on theoretical equations which, in some cases, are simplified to permit convenient use without sacrificing too much accuracy.

Many worked-out examples appear in this book. These examples serve three purposes: acquaint the reader with the use of the formula sets and crosstalk analyses, show the proper use of dimensional quantities such as inches *versus* meters, and relate value magnitudes to physical size. For example, the capacitance between conductors might be in the order of picofarads per inch, rather than microfarads.

All dimensions are assumed to be considerably less than a wavelength. For example, a wavelength at 50 kHz is about $3\frac{3}{4}$ miles, and at 5 MHz, a wavelength is 66 yards, in free space. *Although many of the equations can be applied to transmission lines, the effect of line reflections and other high frequency phenomena are not treated in this book.*

Crosstalk and noise are produced by either voltage drops due to current flow, electromagnetic fields, or both. Some important cases examined herein involve electromagnetic and electrostatic fields. In most examples and experiments, 50 kHz was used for the operating frequency. For specific applications, answers can be obtained by taking note of the frequency dependent terms and scaling accordingly.

The currents are assumed to be flowing on the outermost skin of the conductors so that analogies can be made between inductance, capacitance, and resistance.

Most circuit layout geometries can be considered two-dimensional or at least made up of two-dimensional sections connected together. For example, circuit board lands run parallel or perpendicular to one another on each layer. For this reason, all formula sets present values in capacitance and inductance per unit length, and assume that the conductors have the same cross section and are parallel in the z-axis (into the paper).

In summary, this book introduces or reacquaints the engineer to basic electromagnetic theory. Easy-to-use formulas are presented for determining circuit conductor inductance and capacitance values. Crosstalk causes are reviewed and analyzed. Experiments then demonstrate these principles with measured test data.

Chapter 1
General Concepts

1.1 INTRODUCTION

This chapter provides fundamental background material for general application throughout this book.

Section 1.2 introduces two-dimensional electromagnetic field theory. The close parallels between electric and magnetic fields are described by application of the theory to parallel plate resistors, capacitors, and inductors.

Electric field mapping, discussed in Section 1.3, provides a visual aide to the understanding of electric fields. Applying the principles developed in Section 1.2, this section contains flux plots for several conductor geometries and develops a method for estimating the capacitance directly from the electric field diagrams.

Section 1.4 presents the analogy between self-inductance, capacitance, resistance, and characteristic impedance. This section discusses the relationship of these four quantities and the effects of dielectric medium homogeneity.

1.2 RESISTORS, CAPACITORS, AND INDUCTORS

This section is an introduction to electromagnetic field theory and discusses this subject in simple, yet accurate, easy-to-understand concepts.

Although resistors, capacitors, and inductors have large differences with respect to use and mechanical configuration, they essentially follow the same mathematical rules. Identical equations are used for each, but, as expected, the quantities change. For example, conductance, dielectric constant, and permeability are direct parallels, and each multiplies the appropriate field intensity to get current, electric, and magnetic flux density, respectively.

To illustrate the similarities, an elementary parallel plate geometrical configuration is used. Worked-out examples are presented to give physical meaning to the numbers.

Each subsection ends with a discussion of the assumptions used and directs the reader to additional material found later in this book.

1.2.1 Resistors

Resistors are components widely used in electronic circuit design. In this subsection, the resistor is used to introduce electric field concepts. Moreover, conductance, which is the reciprocal of resistance, is essentially analogous to capacitance. In this subsection we will develop these concepts by starting with a parallel plate resistor shown in Fig. 1.1.

Resistance Defined:

Ohm's law defines resistance as

$$R = \frac{V}{I} \; \Omega \qquad (1)$$

where
R = resistance, ohms (Ω)
V = voltage, volts (V)
I = current, amperes (A)

For the geometry shown in Fig. 1.1, the resistance is

$$R = \frac{\rho d}{wl} \; \Omega \qquad (2)$$

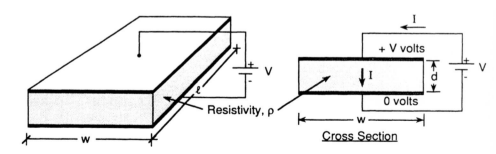

Figure 1.1 A resistor is constructed using two parallel, perfectly conducting plates and a slab of conducting material with resistivity ρ. Voltage source V, applied to the electrodes, creates a current, I, which flows through the resistive material. We assume that negligible current flows through the air outside the resistor.

where

ρ = resistivity of the slab material, Ω-m
d = distance between the plates, m
w = width of the slab, m
l = length of the slab, m

Conductance Defined:

The conductance, G, is

$$G = \frac{I}{V} \text{ S}$$

$$= \frac{1}{R} \tag{3}$$

where

G = conductance, siemens (also called mhos, $\mho$)

For the geometry shown in Fig. 1.1, the conductance is

$$G = \frac{I}{V} = \frac{\sigma w l}{d} \text{ S} \tag{4}$$

where

σ = conductivity, $(\Omega\text{-m})^{-1}$

Example:

Calculate the resistance for the structure shown in Fig. 1.1 with these parameters:

w = 0.1 m (3.9″) l = 0.2 m (7.9″)
d = 0.015 m (0.59″) ρ = 1400 Ω-m

From Eq. (2),

$$R = \frac{\rho d}{wl} \, \Omega$$

$$= \frac{1400 \times 0.015}{0.1 \times 0.2} \, \Omega$$

$$= 1050 \, \Omega \tag{5}$$

Electric Field Intensity, E:

The voltage potential, V, produces an electric field intensity, E, as shown in Fig. 1.2.

The electric field intensity is simply the voltage impressed on the plates divided by distance between them and is given by

$$E = \frac{V}{d} \text{ V/m} \tag{6}$$

Current Flux Density, J:

The electric field intensity, E, causes a current flux density as illustrated in Fig. 1.3. The current flux density is the total current divided by the area of the slab:

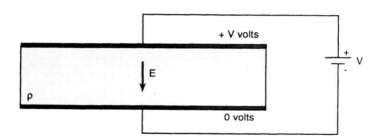

Figure 1.2 An electric field is formed perpendicularly between the plates. Because of the geometrical configuration, this field is uniform throughout the slab.

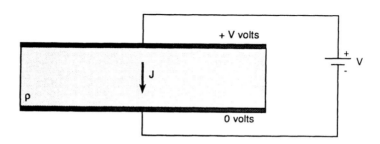

Figure 1.3 The uniform electric field produces a uniform current flux density, J.

$$J = \frac{I}{wl} \text{ A/m}^2 \tag{7}$$

We can find J as a function of E by combining Eqs. (4), (6), and (7):

$$J = \sigma E \text{ A/m}^2 \tag{8}$$

Thus, the current density is equal to the electric field intensity, E, multiplied by the conductivity, σ.

The particular geometry we used produces a uniform electric field intensity. Thus, J and E are scalars. In the more general case, the field is not uniform and the current density can be expressed by vectors (set in boldface characters):

$$\mathbf{J} = \sigma \mathbf{E} \tag{9}$$

Electric Potential, V:

Equipotential surfaces exist between the plates as shown in Fig. 1.4.

Commentary:

1. In the example, we assumed that no current flowed in the air (or vacuum) outside the resistance material. If current did flow, this would be considered fringing current. In most practical cases, fringing current is truly negligible.
2. This subsection illustrates the concept of continuous flux lines. In this case, they are represented by the current. Consider that the total current consists of a number of individual current "lines."

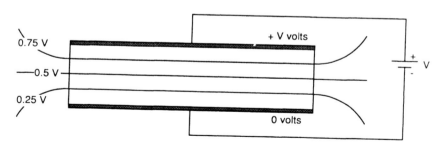

Figure 1.4 Because the electric field intensity, E, is uniform throughout the slab, the equipotential surfaces are evenly spaced between and parallel to the conducting plates. The surfaces are arbitrarily located at increments of 0.25 V, as indicated. These continuous surfaces extend beyond the plates with the curved lines showing the approximate shape near the slab edges.

A current line starts at the potential source, flows through the wire attached to the upper plate, through the resistance material, out the lower plate and back to the negative side of the potential source. The sum of all the individual current lines equals the total current.

1.2.2 Capacitors

This section will follow, step by step, the same outline as Subsection 1.2.1, Resistors. The same figure types and equation formats will be used, in the same order, appropriately modified. In this way, it will become clear that capacitance can be considered to be analogous to conductance. (The $LRCZ_0$ Analogy, Section 1.4, shows that capacitance is analogous to conductance in a number of important situations.)

The classical parallel plate capacitor is shown in Fig. 1.5.

Capacitance Defined:

Capacitance is, by definition:

$$C = \frac{Q}{V} \text{ F} \tag{1}$$

where

    C = capacitance, farads (F)
    Q = charge on one conductor, coulombs (C)
    V = voltage potential between the two conductors, volts (V)

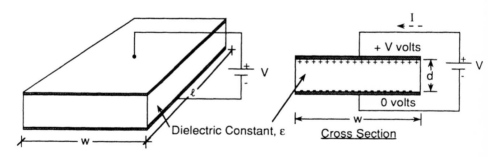

Figure 1.5 A slab of insulating material with dielectric constant ε is contained between two perfectly conducting plates. A voltage applied to the two plates produces an electric field in the dielectric, resulting in polarization. A layer of positive charges (+++) forms on the upper plate/dielectric surface, and an equal number of negative charges (- - -) on the lower interface. The formation of these charges produces a momentary displacement current flow out of the potential source indicated by the dashed current arrow. Please note that *conduction* current *does not* flow through the dielectric (assuming that it is a perfect insulator).

For the geometry shown in Fig. 1.5, the capacitance is

$$C = \frac{\varepsilon w l}{d} \text{ F} \qquad (2)$$

where

ε = dielectric constant of the slab material, F/m
d = distance between the plates, m
w = width of the slab, m
l = length of the slab, m

Alternate Capacitance Definition:

Capacitance can also be defined as

$$C = \frac{\Psi}{V} \text{ F} \qquad (3)$$

where

Ψ = total electric flux produced by potential V, C
V = voltage potential between the two conductors, V

Equation (3) is the result of Gauss's Law for electric fields, as noted in Ref. [1], which states that "the electric flux through any closed surface equals the charge enclosed."

For the geometry shown in Fig. 1.5, the capacitance is

$$C = \frac{\Psi}{V} = \frac{\varepsilon w l}{d} \text{ F} \qquad (4)$$

Example:

Assuming that the fringing flux is negligible, calculate the capacitance for the structure shown in Fig. 1.5 using the same dimensions as the example in Subsection 1.2.1, which are

$w = 0.1$ m (3.9") $\qquad l = 0.2$ m (7.9")

$d = 0.015$ m (0.59") $\qquad \varepsilon = 5 \varepsilon_0 = 5 \times 8.84 \times 10^{-12}$ F/m

From Eq. (2),

$$C = \frac{\varepsilon w l}{d} \text{ F}$$

$$= \frac{5 \times 8.84 \times 10^{-12} \times 0.1 \times 0.2}{0.015} \text{ F}$$

$$= 58.9 \text{ pF} \tag{5}$$

Electric Field Intensity, E:

As with the parallel plate resistor, the voltage potential, V, produces an electric field intensity, E, as shown in Fig. 1.6.

The electric field intensity is the same as for the resistive case:

$$E = \frac{V}{d} \text{ V/m} \tag{6}$$

Electric Flux Density, D:

The electric field intensity, E, causes an electric flux density as illustrated in Fig. 1.7.

The electric flux density is the total electric flux divided by the area of the slab.

$$D = \frac{\Psi}{wl} \text{ C/m}^2 \tag{7}$$

We can find D as a function of E by combining Eqs. (4), (6), and (7):

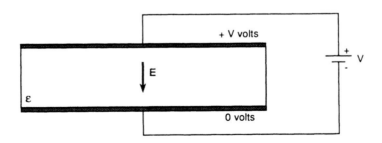

Figure 1.6 An electric field is formed perpendicularly between the plates. Because of the geometrical configuration, this field is uniform throughout the slab.

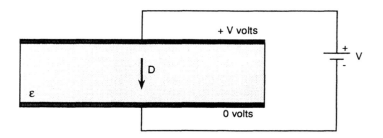

Figure 1.7 The uniform electric field produces a uniform electric flux density, D.

$$D = \varepsilon E \text{ C/m}^2 \tag{8}$$

As stated in words, the electric flux density is equal to the electric field intensity, E, multiplied by the dielectric constant, ε. The particular geometry we used produces a uniform electric field intensity. Thus, D and E may be treated as scalars. In the more general case, the field is not uniform and the electric flux density can be expressed in vector notation:

$$\mathbf{D} = \varepsilon \mathbf{E} \tag{9}$$

Electric Potential, V:

Equipotential surfaces exist between the plates as shown in Fig. 1.8.

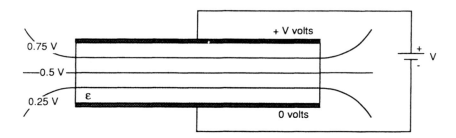

Figure 1.8 Because the electric field intensity, E, is uniform throughout the slab, the equipotential surfaces are evenly spaced between and parallel to the conducting plates. The surfaces are arbitrarily located at increments of 0.25 V, as indicated. These continuous surfaces extend beyond the plates with the curved lines showing the approximate shape near the slab edges.

Commentary:

1. In the above example, we assumed that fringing flux was negligible, and indeed it is for the dimensions shown. However, as expected, this is not always the case. Formula Set C-4 introduces the concept of fringing flux. This shows that the actual capacitance value, as the ratio d/w increases, can be many times greater than would be predicted when fringing is neglected.
2. Electric flux lines in capacitors are continuous in that they originate with the positive charge on one conductor-dielectric interface and terminate with the negative charge on the other.

REFERENCES

1. Kraus, John D., *Electromagnetics*, McGraw-Hill, New York, 1984, p. 40.

1.2.3 Inductors

This subsection will follow the same outline as Subsections 1.2.1 and 1.2.2 and will show another parallel set of equations. Inductance is not analogous to capacitance and conductance in the same way that the latter two are to each other. Electric field equipotential surfaces are replaced by inductive flux lines and electric field flux lines by inductive equipotential surfaces. An inductor consisting of flat, parallel plates appears in Fig. 1.9.

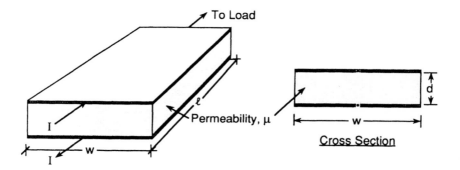

Figure 1.9 A slab of magnetic material with permeability μ is contained between two conducting plates. A current flows down the top plate to the load and returns via the bottom plate. Although this current is represented by two arrows, the current is considered to flow uniformly across the plate surface. This current "sheet" produces a magnetic field in the slab material and in the surrounding medium. In the cross-sectional view, × represents current into the top plate, • is current out of the bottom plate.

Self-inductance Defined:

Self-inductance is, by definition:

$$L = \frac{N\Psi_m}{I} \text{ H} \tag{1}$$

where

L = inductance, henries (H)
Ψ_m = total magnetic flux produced by the current I, webers (Wb)
I = current flowing in the conductor, amperes (A)
N = number of turns (dimensionless)

For the geometry shown in Fig. 1.9, Ref. [2] gives the inductance for $d/w \ll 1$ as:

$$L = \frac{\mu d l}{w} \text{ H} \tag{2}$$

where

μ = permeability of the slab material, H/m
d = distance between the plates, m
w = width of the slab, m
l = length of the slab, m

In this case, the number of turns equals 1. Eq. (1) then becomes

$$L = \frac{\Psi_m}{I} \text{ H} \tag{3}$$

Thus, the inductance for this geometry is

$$L = \frac{\Psi_m}{I} = \frac{\mu d l}{w} \text{ H} \tag{4}$$

Example:

Assuming that the reluctance, defined by Eq. (11), of the medium surrounding the conductors is negligibly small, calculate the inductance for the structure shown in Fig. 1.9 using the same dimensions as the example in Subsection 1.2.1:

$w = 0.1$ m (3.9″) $l = 0.2$ m (7.9″)
$d = 0.015$ m (0.59″) $\mu = \mu_0 = 4\pi \times 10^{-7}$ H/m

From Eq. (2),

$$L = \frac{\mu dl}{w} \text{ H}$$

$$= \frac{4\pi \times 10^{-7} \times 0.015 \times 0.2}{0.1} \text{ H}$$

$$= 0.038 \ \mu\text{H} \tag{5}$$

Magnetic Field Intensity, H:

The current, I, produces a magnetic field intensity as shown in Fig. 1.10. The magnetic field intensity for this specific case is given by

$$H = \frac{I}{w} \text{ A/m} \tag{6}$$

Magnetic Flux Density, B:

The magnetic field intensity, H, causes a magnetic flux density, B, as illustrated in Fig. 1.11.

The magnetic flux density is the total magnetic flux divided by the area of the slab perpendicular to the magnetic field.

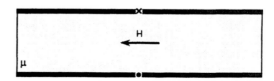

Figure 1.10 A magnetic field is formed parallel to the plates. Because of the geometrical configuration, this field is uniform throughout the slab.

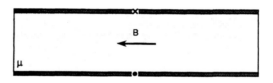

Figure 1.11 The uniform magnetic field produces a uniform magnetic flux density, B.

$$B = \frac{\Psi_m}{dl} \text{ Wb/m}^2 \text{ (also called tesla, T)} \quad (7)$$

We can find B as a function of H by combining Eqs. (4), (6), and (7):

$$B = \mu H \text{ Wb/m}^2 \quad (8)$$

Thus, the magnetic flux density, B, is equal to the magnetic field intensity, H, multiplied by the permeability, $\mu = \mu_r \mu_0$. In vacuum, $\mu = \mu_0 = 4\pi \times 10^{-7}$ H/m. In air, the relative permeability, $\mu_r \approx 1$. For magnetic materials, μ_r can be very much larger than 1 with a corresponding increase in flux density.

The particular geometry that we used would produce a uniform magnetic field intensity. Thus, B and H are treated as scalars. In the more general case, the field is not uniform. The magnetic flux density can be expressed in vector notation as

$$\mathbf{B} = \mu \mathbf{H} \quad (9)$$

Magnetic Potential, U, and Reluctance, $\mathcal{R}$:

Conceptually, the magnetic potential, U, is analogous to electric potential, V.
The magnetic potential difference between two points in the field is given by Ref. [1] as

$$U_1 - U_2 = \int_1^2 \mathbf{H} \cdot dl \, \mathbf{A} \quad (10)$$

As an aid to explaining the following figure, the concept of reluctance, the magnetic circuit equivalent of electrical resistance, is presented, and is given by

$$\mathcal{R} = \frac{U}{\Psi_m} \text{ H}^{-1} \quad (11)$$

In this case, since the magnetic field intensity is uniform, evenly spaced equipotential surfaces lie perpendicular to the plates as shown in Fig. 1.12.

Commentary:

1. The example in this subsection was chosen to illustrate, in the very simplest of terms, magnetic field concepts. The equations are reasonably accurate for the case shown, but are not general in nature. For instance, in the example,

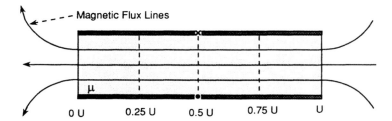

Figure 1.12 Because the magnetic field intensity, H, is uniform throughout the slab, the equipotential surfaces are evenly spaced between and perpendicular to the conducting plates. The surfaces are arbitrarily located at increments of $0.25U$ A as indicated. These surfaces stop at the right-hand edge of the plate because the reluctance, $\mathcal{R}$, of the magnetic circuit formed by the medium outside the conductors is assumed to be approximately zero. This is because the volume of the surrounding medium is very large compared to the volume between the plates. Continuous magnetic flux lines extend beyond the plate edges, with individual lines leaving the left edge, traveling through the medium surrounding the plates, and returning to the right edge.

the spacing-to-width ratio, d/w, for the parallel conductors was made small so that the reluctance of the medium surrounding the conductors would be small compared to that between the plates. Subsection 1.2.1 assumed that no current flowed outside the resistive material; Subsection 1.2.2 assumed that no electric flux lines were outside of the dielectric; here, in an analogous way, the magnetic equipotential surfaces stop at the right-hand edge of the conductors at potential U. Formula Set L-4 includes the concept of the fringing factor, K_{L1}, which is, in the inductance case, magnetic potential fringing. As the ratio d/w increases, the magnetic potential also increases. The inductance decreases because the reluctance of the magnetic circuit outside the plates has increased.

2. Magnetic flux lines are continuous. The magnetic flux, Ψ_m, leaves the left-hand edge of the magnetic circuit, as shown in Fig. 1.12, travels through the surrounding medium and returns via the right-hand edge. Because the reluctance is assumed to be zero, Eq. (11) states that the magnetic potential drop will also be zero for this path. Thus, the magnetic potential remains at $\approx U$. Please refer to texts on electromagnetic field theory for further information regarding magnetic potential.

REFERENCES

1. Kraus, John D., *Electromagnetics*, New York, McGraw-Hill, 1984, p. 173.
2. Zahn, Markus, *Electromagnetic Field Theory*, Robert K. Krieger, Malabar, Florida, 1979, p. 570.

1.3 ELECTRIC FIELD MAPPING

The mapping of electric fields, shown by flux plots, provides valuable insight into the nature of electromagnetic fields. This section describes the theory of flux plotting and provides examples for several geometries. Capacitance values estimated directly from the field plots are compared with values predicted in Subsection 1.2.2 and formula sets appearing in Chapter 2.

1.3.1 Parallel Plate Capacitor

Flux tubes and equipotential surfaces form the basis for electric field mapping. Fig. 1.13 shows a parallel plate capacitor similar to that shown in Subsection 1.2.2.

The capacitance for each unit cell can be determined from Subsection 1.2.2, Eq. (2):

$$C = \frac{\varepsilon w l}{d} \text{ F} \qquad (1)$$

For the unit cell:

$d = \Delta d$ = distance between unit cell top and bottom, m
$w = \Delta w$ = width of the cell, m

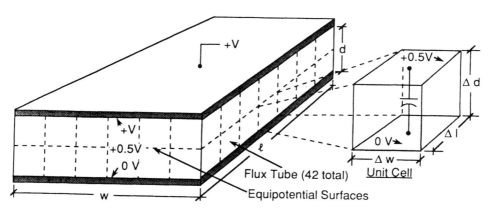

Figure 1.13 This capacitor representation contains three equipotential surfaces at 0 V, 0.5 V, and 1 V. Each of the forty-two flux tubes carries the same number of flux lines and has two unit cells in series. These cells are cubes with equal face dimensions. One of the flux cells is enlarged as shown on the right. This may be considered as a unit capacitor because it is "connected" to two equipotential surfaces.

$l = \Delta l$ = length of the cell, m
ε = material dielectric constant, F/m

Substituting these quantities in Eq. (1) yields the capacitance, C_c, for the unit cell:

$$C_c = \frac{\varepsilon \Delta w \Delta l}{\Delta d} \text{ F} \tag{2}$$

If we let $\Delta d = \Delta w$ as is shown in Fig. 1.13, the value of C_c is equal to $\varepsilon \Delta l$ or

$$C_c = \varepsilon \Delta l \text{ F} \tag{3}$$

If we let $\Delta l = \Delta w$, we can solve for the unit capacitor value but we must specify the value of Δl because ε has dimensions in F/m. For example, if Δl is 1m and $\varepsilon = \varepsilon_0 = 8.84 \times 10^{-12}$ F/m, the capacitance of a unit cell, 1 m on the side, is 8.84 pF. If $\Delta l = 1$ cm (0.01 m), the unit cell capacitance would then be 0.0884 pF. Because, on the front face, there are n_p unit capacitors in parallel and n_s in series (in this case 6 and 2, respectively), the total capacitance is

$$C = \frac{n_p}{n_s} C_c \tag{4}$$

$$= \frac{\varepsilon n_p \Delta l}{n_s} \text{ F} \tag{5}$$

Letting $\Delta l = l$, the capacitance per unit length is then given by

$$\frac{C}{l} = \varepsilon \frac{n_p}{n_s} \text{ F/m} \tag{6}$$

Thus, the capacitance per unit length is the dielectric constant times the number of unit cells in parallel divided by the number in series. It is important to emphasize that *the unit cells have square faces*, i.e., $\Delta w = \Delta d$.

Example 1:

Using Eq. (6), calculate the capacitance for the structure shown in the example in Subsection 1.2.2. The dimensions are

$w = 0.1$ m (3.9") $\qquad l = 0.2$ m (7.9")

$d = 0.015$ m (0.59") $\qquad \varepsilon = 5\varepsilon_0 = 5 \times 8.84 \times 10^{-12}$ F/m

The capacitor is divided into squares as shown in Fig. 1.14:
The capacitance per unit length is

$$\frac{C}{l} = \varepsilon \frac{n_p}{n_s} \text{ F}$$

$$= 5 \times 8.84 \times 10^{-12} \left(\frac{20}{3}\right) \text{ F/m}$$

$$= 295 \text{ pF/m} \tag{7}$$

Because $l = 0.2$ m, the capacitance is

$$C = 0.2 \times 295$$
$$= 58.9 \text{ pF} \tag{8}$$

which is the same value we got in Subsection 1.2.2.

1.3.2 Circular Conductor Capacitor

Circles are used to generate flux plots for circular conductors as illustrated in Fig. 1.15. (Note that this flux plot was drawn by using equations developed in Appendix A.)

For the parallel plate capacitor, the unit cells had square faces. In this case, the cells are curvilinear squares. In Fig. 1.15, while the cells near the origin are almost square, those farther out become quite distorted. Cell A is a good example but, even here, cell A can be sub-divided, as sketched in Fig. 1.15, into smaller cells which again approach square shapes.

Flux plots require these conditions:
1. Equipotential surfaces are continuous.

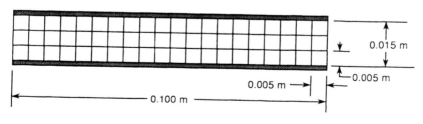

Figure 1.14 In this example, each square is 0.005 m on the side. Thus, $n_p = 0.100/0.005 = 20$ and $n_s = 0.015/0.005 = 3$.

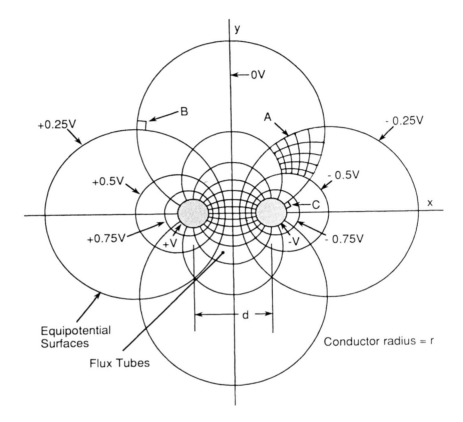

Figure 1.15 Two long, circular conductors, each having a radius $r = 2$ units, and spaced $d = 10$ units apart are shown in this cross-sectional view. The conductors are at potentials $+V$ and $-V$ respectively. Equipotential surfaces are cylinders and at potentials as shown. Flux tubes are formed by circular boundaries. The ratio d/r has been deliberately selected to yield an integral number of flux tubes. Please note that this figure is symmetrical about both the x and y axes. When two-axis symmetry is present, only one quadrant need be drawn to get the desired field information. (This flux plot was drawn using equations developed in Appendix 1.)

2. Flux tube boundaries and equipotential surfaces are orthogonal, that is, cross at right angles. Please note point B in Fig. 1.15.
3. Flux tube boundaries start and terminate on free charges on conductors and are perpendicular to the conductors at these points (point C is an example).
4. For axis symmetry, as in Fig. 1.15, the y axis is a zero potential surface. The x axis is a flux tube boundary.

Example 2:

Calculate the capacitance per unit length between the two conductors shown in Fig. 1.15. Compare this value with that given by Formula Set C-1, Eq. (1).
The capacitance per unit length given by Eq. (6) is

$$\frac{C}{l} = \varepsilon \frac{n_p}{n_s} \text{ F/m}$$

In this case, the capacitance consists of 16 unit capacitors in parallel and 8 in series corresponding to the 16 parallel flux tubes and 8 equipotential intervals. Hence, $n_p = 16$ and $n_s = 8$. Thus,

$$\frac{C}{l} = \varepsilon \frac{16}{8} \text{ F/m}$$

$$= 2\varepsilon \text{ F/m} \tag{9}$$

Formula Set C-1, Eq. (1) gives the capacitance per unit length for long circular conductors as

$$\frac{C}{l} = \frac{\pi \varepsilon_r \varepsilon_0}{\ln\left[\frac{d}{2r} + \sqrt{\left(\frac{d}{2r}\right)^2 - 1}\right]} \text{ F/m} \tag{10}$$

For $d = 10$ units and $r = 2$ units,

$$\frac{C}{l} = \frac{\pi \varepsilon_r \varepsilon_0}{\ln\left[\frac{10}{2 \times 2} + \sqrt{\left(\frac{10}{2 \times 2}\right)^2 - 1}\right]} \text{ F/m}$$

$$= 2.005 \, \varepsilon \text{ F/m} \tag{11}$$

(To get exactly 16 flux tubes with 8 equipotential surfaces, the radii of the circular conductors have to be 1.9928 ... units. Substituting this value in Eq. (11) yields $C/l = 2.000081$.)

1.3.3 Elliptical-Rectangular Conductor Capacitors

Formula Set C-5 gives the capacitance per unit length for long, parallel, rectangular conductors, and the derivation is based on the premise that flux patterns will be approximately the same for rectangular and circular conductors of equivalent areas.

Figure 1.16 uses a flux plot to illustrate this point. The circular conductors are changed to elliptical conductors with equivalent perimeter lengths.

1.3.4 Rectangular Conductor between Two Ground Planes

Formula Set C-9 gives the capacitance per unit length between a rectangular conductor and two ground planes with the latter connected together.

Example 3:

Determine the capacitance per unit length for the configuration shown in Fig. 1.17 and compare this value with that calculated using Formula Set C-9.

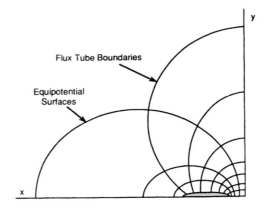

Figure 1.16 This flux plot shows the one-quadrant approximate field distribution for two elliptical conductors. The plot contains the same number of equipotential surfaces and flux tubes suggesting that the capacitance per unit length is the same as for the circular conductors shown in Fig. 1.15. Close examination shows that the plot is not perfect: some of the crossing angles are not exactly 90°, some flux tube boundaries do not terminate exactly perpendicular to the conductor and the curvilinear squares are not ideal.

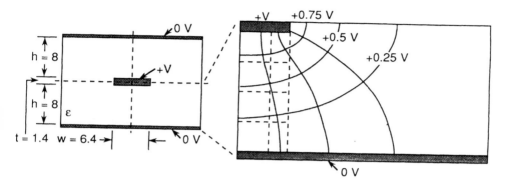

Figure 1.17 A rectangular conductor is located midway between two ground planes and has the dimensions shown in the left-hand figure. An enlarged view of the lower right quadrant appears on the right with sketched flux tube boundaries and equipotential surfaces.

From the sketch, there are 4 equipotential intervals and 16 flux tubes, 4 for each quadrant. Thus, from Eq. (6),

$$\frac{C}{l} = \varepsilon \frac{n_p}{n_s} \text{ F/m}$$

$$= \varepsilon \frac{16}{4} \text{ F/m}$$

$$= 4\varepsilon \text{ F/m} \qquad (12)$$

From Formula Set C-9, the capacitance per unit length is given approximately as

$$\frac{C}{l} = 2\varepsilon_r \varepsilon_0 K_{C2}\left(\frac{w}{h}\right) \text{F/m} \qquad (13)$$

where

ε_r = relative dielectric constant, F/m
ε_0 = dielectric constant of free space, F/m
K_{C2} = fringing factor, dimensionless
w = conductor width, m
h = distance above the ground plane, m

Formula Set C-9 gives $K_{C2} = 2.4$ for the dimensions shown in Fig. 1.17. Because $\varepsilon_r \varepsilon_0 = \varepsilon$, Eq. (14) gives C/l as

$$\frac{C}{l} = 2\varepsilon_r\varepsilon_0 K_{C2}\left(\frac{w}{h}\right) \text{F/m}$$

$$\frac{C}{l} = 2 \times 2.4 \times \left(\frac{6.4}{8}\right)\varepsilon \text{ F/m}$$

$$= 3.84\ \varepsilon \text{ F/m} \quad (14)$$

Commentary and Conclusions:

1. Flux plotting can provide, with simple sketches, rough order-of-magnitude capacitance (and hence resistance and inductance) values for geometries not amenable to simple analytical solutions.

    However, high-level accuracy plots drawn by hand are tedious and time-consuming. For this reason, finite element analysis routines have been developed which can produce computer-generated plots such as those shown in Refs. [1] and [2].

REFERENCES

1. Hoole, S. and Ratnajeevan, H., *Computer-Aided Analysis and Design of Electromagnetic Devices* New York, Elsevier, 1989.
2. Pantic, Z. and Mittra, R., "Quasi-TEM Analysis of Microwave Transmission Lines by the Finite Element Method," *IEEE Trans.*, Vol. MTT-34, No. 11, November 1986.

1.4 THE $LCRZ_0$ ANALOGY*

The analogy between self-inductance, capacitance, resistance, and characteristic impedance is useful in the derivation of equations for given geometrical conductor arrangements. References often give one only of these qualities, for example, the characteristic impedance. For *homogeneous media,* when one of the quantities is known, the other three can be determined by use of a geometrical factor, Γ. If the medium is *nonhomogeneous,* we need to know *two* (or three) of the quantities to get the other two (or one). Therefore, there are two cases:

1. The conductors are completely surrounded by a medium which is homogeneous This means that the medium is the same everywhere, except that of the volume

---

*Our discussion of the geometrical factor concept, which is accurate with respect to conductors in homogeneous media, is largely based on Catt, I., *Digital Hardware Design* (London, Macmillan 1979), pp. 7–11.

actually occupied by the conductors. Thus, *surrounded* means between the conductors and outside the conductors.
2. The surrounding medium is nonhomogeneous. Please see the discussion on applications following the example. (In all cases, the surrounding medium is assumed to be isotropic and linear, i.e., the properties are independent of applied field direction or magnitude.)

Equations:

Case 1—Homogeneous Surrounding Medium: Reference [1] describes the $LRCZ_0$ analogy principle in easy-to-understand terms and Ref. [2] provides theoretical justification. Following the approach of Ref. [1], adapted for this book, the resistance, conductance, capacitance, inductance, and characteristic impedance can be expressed in terms of the material properties and the geometrical factor, Γ.

Resistance:

$$Rl = \frac{\rho}{\Gamma} \Omega\text{-m} \qquad (1)$$

where

R = resistance, Ω
l = length of the resistor, m
ρ = resistivity, Ω-m
Γ = geometrical factor (dimensionless)

Conductance:

$$\frac{G}{l} = \sigma \Gamma \text{ S/m} \qquad (2)$$

where

G = conductance, S (or mhos)
σ = conductivity, S/m

Capacitance:

$$\frac{C}{l} = \varepsilon \Gamma \text{ F/m} \qquad (3)$$

where

C = capacitance, F
ε = dielectric constant, F/m

Self-Inductance:

$$\frac{L}{l} = \frac{\mu}{\Gamma} \text{ H/m} \tag{4}$$

where

L = self-inductance, H
μ = permeability, H/m

Characteristic Impedance:

$$Z_0 = \frac{1}{\Gamma}\sqrt{\frac{\mu}{\varepsilon}} \; \Omega \tag{5}$$

where

Z_0 = characteristic impedance, Ω (for lossless configurations).

Rearranging and combining Eqs. (3), (4), and (5) yields these important relationships:

$$\Gamma = \frac{\left(\frac{C}{l}\right)}{\varepsilon} = \frac{\mu}{\left(\frac{L}{l}\right)} = \frac{1}{Z_0}\sqrt{\frac{\mu}{\varepsilon}} \tag{6}$$

$$\frac{L}{l} = \frac{\mu\varepsilon}{\left(\frac{C}{l}\text{ F/m}\right)} \text{ H/m} \tag{7}$$

$$= Z_0 \Omega \sqrt{\mu\varepsilon} \text{ H/m} \tag{8}$$

$$\frac{C}{l} = \frac{\mu\varepsilon}{\left(\frac{L}{l}\text{ H/m}\right)} \text{ F/m} \tag{9}$$

$$= \frac{\sqrt{\mu\varepsilon}}{Z_0 \Omega} \text{ F/m} \tag{10}$$

$$\left(\frac{L}{l} \text{ H/m}\right)\left(\frac{C}{l} \text{ F/m}\right) = \mu\varepsilon \text{ H/m F/m} \tag{11}$$

$$Z_0 = \frac{\left(\frac{L}{l} \text{ H/m}\right)}{\sqrt{\mu\varepsilon}} \Omega \tag{12}$$

$$= \frac{\sqrt{\mu\varepsilon}}{\left(\frac{C}{l} \text{ F/m}\right)} \Omega \tag{13}$$

$$Z_0 = \sqrt{\frac{L/l \text{ H/m}}{C/l \text{ F/m}}} \Omega \tag{14}$$

$$= \sqrt{\frac{L}{C}} \Omega \tag{15}$$

Case 2—Surrounding Medium is Nonhomogeneous: If the medium surrounding the conductors is nonhomogeneous, Eqs. (1) through (13) are modified by fringing factors. Equations (1) through (5) are thus revised as follows.

Resistance:

$$Rl = \frac{\rho}{K_R \Gamma} \Omega \tag{16}$$

Conductance:

$$\frac{G}{l} = \sigma K_R \Gamma \text{ S/m} \tag{17}$$

Capacitance:

$$\frac{C}{l} = \varepsilon K_C \Gamma \text{ F/m} \tag{18}$$

Self-Inductance:

$$\frac{L}{l} = \frac{\mu}{K_L \Gamma} \text{ H/m} \tag{19}$$

Characteristic Impedance:

$$Z_0 = \frac{1}{\sqrt{K_L K_C}\, \Gamma} \sqrt{\frac{\mu}{\varepsilon}}\, \Omega \tag{20}$$

where

K_R = Resistive fringing factor (dimensionless)
K_C = Capacitive fringing factor (dimensionless)
K_L = Inductive fringing factor (dimensionless)
Γ = Geometrical factor (same as used in Eqs. (1) through (6)).

Example:

We will first determine the geometrical factor for the parallel plate resistor with the structure shown in Fig. 1.1, Subsection 1.2.1. Strictly speaking, the medium surrounding the conductors is not homogeneous because, in the case of the resistor and capacitor, the material between the plates is different than that outside the plates. However, the dimensions have been chosen so that fringing is negligible meaning that K_L and K_C are nearly equal to 1.

The example has these parameters:

$w = 0.1$ m (3.9") $\qquad l = 0.2$ m (7.9")

$d = 0.015$ m (0.59") $\qquad \rho = 1400$ Ω-m

Using the geometrical factor for this configuration, calculate the resistance, capacitance and self-inductance for the examples shown in Subsections 1.2.1, 1.2.2 and 1.2.3 respectively.

From Eq. (2), Subsection 1.2.1,

$$R = \frac{\rho d}{wl}\, \Omega \tag{21}$$

or

$$Rl = \frac{\rho d}{w}\, \Omega/\text{m} \tag{22}$$

Equation (1) is

$$Rl = \frac{\rho}{\Gamma} \Omega$$

By inspection,

$$\Gamma = \frac{w}{d} \text{ (dimensionless)} \qquad (23)$$

Using the dimensions from the example,

$$\Gamma = \frac{0.1 \text{ m}}{0.015 \text{ m}}$$
$$= 6.66\ldots \qquad (24)$$

The resistance is then from Eq. (1):

$$R = \frac{\rho}{l\Gamma} \Omega$$
$$= \frac{1400 \ \Omega\text{-m}}{0.2 \text{ m} \times 6.66\ldots} \Omega$$
$$= 1050 \ \Omega \qquad (25)$$

which is what we got in Subsection 1.2.1. The example in Subsection 1.2.2 showed a parallel plate capacitor with the same dimensions as the parallel plate resistor. The calculated value was 58.9 pF using a dielectric constant of $5 \times 8.84 \times 10^{-12}$ F/m. From Eq. (3), the capacitance per unit length is

$$\frac{C}{l} = \varepsilon \Gamma \text{ F/m}$$

The capacitance is then:

$$C = \varepsilon l \Gamma \text{ F}$$
$$= 5 \times 8.84 \times 10^{-12} \times 0.2 \text{ m} \times 6.66\ldots \text{ F}$$
$$= 58.9 \text{ pF} \qquad (26)$$

The example in Subsection 1.2.3 shows a parallel plate inductor with a calculated self-inductance of 0.038 μH with $\mu = \mu_0 = 4\pi \times 10^{-7}$. Equation (4) gives the self-inductance per unit length as

$$\frac{L}{l} = \frac{\mu}{\Gamma} \text{H/m}$$

or

$$L = \frac{\mu l}{\Gamma} \text{H} \tag{27}$$

Substituting the values:

$$L = \frac{4\pi \times 10^{-7} \times 0.2 \text{ m}}{6.66 \ldots} \text{H/m}$$

$$= 0.038 \text{ mH} \tag{28}$$

LCRZ$_0$ Analogy Applications:

As noted in the introduction to this section, the *LCRZ$_0$* analogy depends on the surrounding medium being homogeneous.

For practical purposes, these important geometries can meet the homogeneity requirement:

1. Parallel conductors in free space (Formula Sets C-1, L-1).
   - Bare conductors assumed.
2. Stripline geometries (Formula Sets C-9, L-9).
   - Conductors-to-ground plane dimensions small compared to conductor-to-structure-discontinuity distances, such as ground plane edges, adjacent parallel lands, and so forth.
3. Coaxial cables (Formula Sets C-11, L-11).
   - Dielectric separating conductors is homogeneous.

When using the *LRCZ$_0$* Analogy, each of the following geometries should be reviewed for the degree of surrounding medium homogeneity.

1. Circular insulated conductors (Formula Sets C-1, L-1).
   - C/l depends on $\varepsilon_{r(\text{eff})}$, L/l does not.
2. Circular insulated conductors near a ground plane (Formula Sets C-2, L-2).
   - (Same comments as 1.)
3. Vertical flat conductors (Formula Sets C-4, L-4).
   - Fringing factors K_{L1} and K_{C1} provide the necessary adjustment for nonhomogeneous medium.

4. Horizontal flat conductors (Formula Sets C-5, L-5).
   - (Same comments as 1.)
5. Flat conductors near a ground plane (Formula Sets C-6, L-6).
   - (Same comments as 3.)

Commentary and Conclusions:

1. For the particular dimensions used in the example, we have seen that the geometrical factor concept is valid.
2. Ref. [2] shows that this concept is valid for all parallel conductors provided that the following conditions are met:
   - The region between and surrounding the electrodes is homogeneous, with no free charge and is current free;
   - The electrodes are highly conducting;
   - Currents flow on the conductor outer surfaces, a good approximation, especially at higher frequencies where the current depth of penetration is small compared with the conductor dimensions. Please see Subsection 5.2.2 for further details. Thus, the self-inductance internal to the conductors is neglected. (Reference [2] uses thin-walled hollow conductors);
   - The geometry is two-dimensional, meaning that the conductors are parallel and long with respect to their other dimensions.

   These conditions are generally compatible with the assumptions made and accuracy levels used in this book with the preceding limitations noted.
3. For a given conductor geometry, the self-inductance between conductors depends only on the relative permeability, μ_r, (usually = 1), and is independent of the dielectric constant. Thus, when calculating inductance from capacitance or characteristic impedance, the effective relative dielectric constant, $\varepsilon_{r(eff)}$, must be used. Formula Set L-1 discusses this in further detail.
4. Characteristic impedance values are readily available with dimensional ratio limitations usually stated. Reference [3] shows a number of different geometrical relationships.

REFERENCES

1. Catt, I., *Digital Hardware Design*, London, Macmillan, 1979, pp. 7–8.
2. Zahn, Markus, *Electromagnetic Field Theory: A Problem Solving Approach*, Malabar, Florida, Robert E. Krieger, pp. 458–459.
3. *Reference Data For Radio Engineers*, Indianapolis, Howard W. Sams, 1968.

Chapter 2
Formula Sets

2.1 INTRODUCTION

This chapter contains 25 formula sets covering circuit geometries commonly used in printed wiring board and ceramic module design as well as other areas of electronic engineering. A brief description of the format used follows.

Numbering:

For convenience, corresponding capacitive and inductive formula sets, with the same geometry, carry the same number. For example, Formula Sets C-5 and L-5 describe the equations for horizontal flat conductors. Two exceptions are C-12 for spheres and L-12 for loops. The subsection numbers apply as well. C-5 appears in Subsection 2.2.5, and L-5 in 2.3.5. The equations for characteristic impedance are collected in one formula set, Z_0-(ALL).

Equations (Examples):

Each equation set is expressed in both SI and English units with the Formula Set C-1 as an example:

$$\frac{C}{l} \approx \frac{\pi \varepsilon_{r(\text{eff})} \varepsilon_0}{\ln\left(\frac{d}{r}\right)} \text{ F/m}, \quad \text{for } \frac{2r}{d} \ll 1 \qquad (1)$$

$$= \frac{27.8 \, \varepsilon_{r(\text{eff})}}{\ln\left(\frac{d}{r}\right)} \text{ pF/m} \qquad (2a)$$

$$= \frac{0.7\varepsilon_{r(\text{eff})}}{\ln\left(\dfrac{d}{r}\right)} \text{ pF/in} \qquad (2b)$$

SI units are used in Eqs. (1) and (2a); English units (2b). The suffix a is for SI units, and b is for English units. Designations for dimensions are consistent throughout this book. For example, d = distance, r = radius, h = height, and l = length.

Measurements:

Most formula sets include test data supporting the equations. The examples show predicted values, based on the formula set equations, which are then compared with measured test data. Effects of conductor end fringing are included in the data, but not in the equations. Thus, capacitive measurements tend to be slightly larger than predicted.

Similarly, for self-inductance loops, the equations do not allow for far-end connections. In this case, the measured values are adjusted to account for the inductance added by these connections.

Derivations:

For convenience, derivations are included in the formula set if applicable. Equations "borrowed" from other sections are repeated.

Commentary and Conclusions:

Following the end of each formula set, as appropriate, is a brief discussion covering issues related to the topic.

2.2 CAPACITANCE

2.2.1 Capacitance between Two Circular Conductors, C-1

The capacitance between circular conductors often determines the cut-off frequency in circuits using twisted pairs or flat cable. This formula set provides the exact and approximate equations for determining the conductor-to-conductor capacitance and compares values with measured results. The effect of wire insulation is evaluated.

Reference [1] gives the exact expression for the capacitance per unit length between the conductors shown in Fig. 2.1.

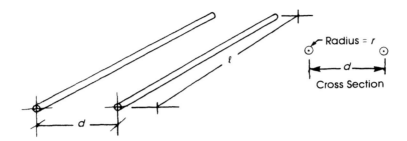

Figure 2.1 Two long, circular conductors of radius r are separated by distance d. Wire insulation is not shown.

Equations:

$$\frac{C}{l} = \frac{\pi \varepsilon_r \varepsilon_0}{\ln\left[\frac{d}{2r} + \sqrt{\left(\frac{d}{2r}\right)^2 - 1}\right]} \text{ F/m}$$

$$= \frac{\pi \varepsilon_r \varepsilon_0}{\ln\left\{\frac{d}{2r}\left[1 + \sqrt{1 - \left(\frac{2r}{d}\right)^2}\right]\right\}} \text{ F/m} \qquad (1)*$$

$$\frac{C}{l} \approx \frac{\pi \varepsilon_{r(\text{eff})} \varepsilon_0}{\ln\left(\frac{d}{r}\right)} \text{ F/m}, \qquad \text{for } \frac{2r}{d} \ll 1 \qquad (2)$$

$$= \frac{27.8 \; \varepsilon_{r(\text{eff})}}{\ln\left(\frac{d}{r}\right)} \text{ pF/m} \qquad (3a)$$

$$= \frac{0.7 \; \varepsilon_{r(\text{eff})}}{\ln\left(\frac{d}{r}\right)} \text{ pF/in} \qquad (3b)$$

Note: Reference [1] gives Eq. (1) in terms of h and R which are changed to d and r respectively, to be consistent with nomenclature used in this book. Similarly, D, a_1, and a_2 in Eq. (4) are changed to d, r_1, and r_2.

where

ε = dielectric constant, F/m
ε_0 = dielectric constant in vacuum
$= \dfrac{10^{-9}}{36\pi}$ F/m
$= 8.84 \times 10^{-12}$ F/m
$\varepsilon_{r(\text{eff})} \equiv$ effective dielectric constant accounts for nonhomogeneity of the region surrounding conductors

For conductors of unequal radii, Ref. [1] gives

$$\frac{C}{l} \approx \frac{2\pi\varepsilon}{\ln(d^2/r_1 r_2)} \text{ F/m} \qquad (4)$$

Example 1:

Calculate the capacitance per foot between two adjacent conductors for the flat cable shown in Fig. 2.2. Using Eq. (3b), and letting $\varepsilon_r = 1.0$ and 3.2, we get

$$\frac{C}{l} = \frac{0.7\,\varepsilon_r}{\ln\left(\dfrac{0.050''}{0.0075''}\right)} \times 12'' \text{ pF/ft}$$

$= 4.4$ pF/ft ($\varepsilon_r = 1.0$)

$= 14.2$ pF/ft ($\varepsilon_r = 3.2$) $\qquad (5)$

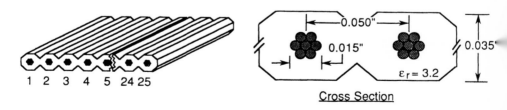

Cross Section

Figure 2.2 Flat cable made up of parallel #28 stranded conductors on 0.050" centers with *polyvinyl chloride* (PVC) insulation ($\varepsilon_r \approx 3.2$).

The value measured for a sample of this cable was 9.4 pF/foot which would be the result if $\varepsilon_{r(\text{eff})} = 2.1$. This is expected because the PVC insulation only partially surrounds the conductors. The effective dielectric constant, $\varepsilon_{r(\text{eff})}$, is thus expected to be in the range $1 \leq \varepsilon_{r(\text{eff})} \leq 3.2$. Table 2.1 lists measurements made at $f = 10$ kHz for various flat conductor pairs. The measured values are compared with values calculated using $\varepsilon_r = 1$. The effective dielectric constant, $\varepsilon_{r(\text{eff})}$, is then determined.

The results shown in Table 2.1 indicate, as expected, that the effective relative dielectric constant, $\varepsilon_{r(\text{eff})}$, decreases with increasing distance. This is because proportionally more of the electric flux tubes are in the air where $\varepsilon_r \approx 1$ than in the PVC insulation. These measurements were made in free air and include, in their values, effects due to other conductors in the cable.

Note: Flat cable capacitances are measured using special test techniques and catalog values are listed accordingly. Please refer to the manufacturers' catalogs for details.

Example 2:

Determine the capacitance between two conductors, each with $r = 0.0075''$, in a twisted pair 1 foot long where the PVC insulation thickness equals the conductor radius. Compare the values obtained using both Eqs. (1) and (3b).

In this case, $d = 4r = 4 \times 0.0075'' = 0.030''$ with $r = 0.0075''$ as before. Using Eq. (1), modified for English units, we get

Table 2.1
Capacitance between Various Conductor Pairs in a Flat Cable

Pair	Measured (pF/ft)	Calculated, $\varepsilon_r = 1$ (pF/ft)	$\varepsilon_{r(\text{eff})}$
1-2	9.4	4.4	2.1
1-3	6.2	3.2	1.9
1-4	4.7	2.8	1.7
1-5	3.9	2.6	1.5
1-6	3.4	2.4	1.4
1-7	3.0	2.3	1.3

$$C = \frac{0.7 \times 3.2 \times 12''}{\ln\left\{\dfrac{0.030''}{0.015''}\left[1 + \sqrt{1 - \left(\dfrac{0.015''}{0.030''}\right)^2}\right]\right\}}$$

$$= 20.4 \text{ pF} \tag{6}$$

Using the approximate expression given by Eq. (3b),

$$C \approx \frac{0.7 \times 3.2 \times 12''}{\ln\left(\dfrac{0.030''}{0.0075''}\right)}$$

$$= 19.4 \text{ pF} \tag{7}$$

Thus, for this example, there is only a 5% difference between using Eqs. (1) and (3b). (Because the conductors are not fully embedded in PVC, the capacitance will be smaller than that indicated by Eqs. (6) and (7).)

Commentary and Conclusions:

1. The capacitance per unit length is independent of physical size so long as the ratio d/r remains constant and the effect of $\varepsilon_{r(\text{eff})}$ is properly taken into account.

REFERENCES

1. Johnk, C.T.A., *Engineering Electromagnetic Fields and Waves*, New York, John Wiley and Sons, 1973, pp. 232, 234.

2.2.2 Capacitance between a Circular Conductor and a Ground Plane, C-2

Occasionally, a single circular conductor is used near a conducting ground plane. The capacitance per unit length between a long circular conductor and a conducting ground plane shown in Fig. 2.3 is given by Eq. (1).

Equations:

The capacitance per unit length for $h/r \gg 1$ is given approximately by

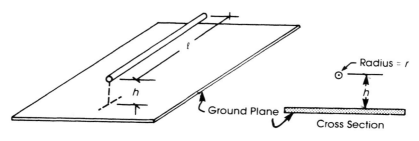

Figure 2.3 A long, circular conductor spaced distance h over a ground plane. The return circuit is via the ground plane.

$$\frac{C}{l} \approx \frac{2\pi\varepsilon_r\varepsilon_0}{\ln\left(\frac{2h}{r}\right)} \text{ F/m} \tag{1}$$

$$= \frac{55.6\varepsilon_{r(\text{eff})}}{\ln\left(\frac{2h}{r}\right)} \text{ pF/m} \tag{2a}$$

$$= \frac{1.4\varepsilon_{r(\text{eff})}}{\ln\left(\frac{2h}{r}\right)} \text{ pF/in} \tag{2b}$$

Example:

Calculate the capacitance between 1 ft of a flat cable conductor and a ground plane placed immediately below the cable as shown in Fig. 2.4.

Using Eq. (2b),

$$\frac{C}{l} = \frac{1.4 \times 1.0}{\ln\left(\frac{2 \times 0.0175''}{0.0075''}\right)} \times 12'' \text{ pF}$$

$$= 10.9 \text{ pF/ft (for } \varepsilon_r = 1) \tag{3}$$

For a 1' cable sample, the measured value was 22.6 pF yielding an effective dielectric constant, $\varepsilon_{r(\text{eff})} = 2.1$, showing that, proportionally, about the same number of electric flux tubes are in the PVC as were for the adjacent conductors measured in Formula Set C-1.

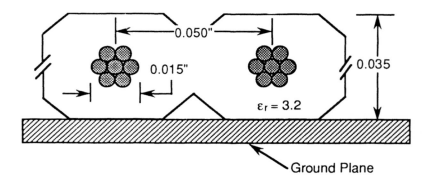

Figure 2.4 This is the same flat cable used in Formula Set C-1, Example 1. Here the #28 stranded conductors are centered 0.0175" above the ground plane, and separated by the cable PVC insulation. Conductor 12 (of a 25-conductor cable) was used in the measurement.

Commentary:

1. Note that the capacitance calculated above is exactly twice that for two parallel conductors if the spacing $d = 2h$. This is because, in the case of the parallel conductors, one at $+V$ volts, the other at $-V$ volts, an equipotential plane at zero volts lies halfway between. This plane can be considered the equivalent to a ground plane. Subsection 1.2.2 defines capacitance as

$$C = \frac{Q}{V}$$

The total voltage difference for the two conductors case is $2V$; for the single conductor and the ground plane, it is V. The conductor charge is the same for both cases. Thus,

$$C = \frac{Q}{V} \quad \text{for a single conductor and the ground plane,}$$

$$C = \frac{Q}{2V} \quad \text{for two conductors.}$$

2. The comment for Formula Set C-1 with regard to relative geometry applies equally to this case as do the equation approximations.

2.2.3 Two-Circular-Conductor Mutual Capacitance Near a Ground Plane, C-3

Ground planes can produce very substantial reductions in crosstalk between circuits due to the decreased mutual capacitance.

This formula set is for circular conductors (wires) running close to a ground plane such as a metal chassis. More importantly, this formula set introduces the concept of mutual capacitance and its relationship to mutual inductance and characteristic impedance.

Please see Formula Set C-7 for printed wiring board applications.

Equations:

The mutual capacitance per unit length for the conductors shown in Fig. 2.5 is given approximately as

$$\frac{C_m}{l} \approx \frac{\pi \varepsilon_{r(\text{eff})} \varepsilon_0 \ln\left[1 + \left(\frac{2h}{d}\right)^2\right]}{\ln\left(\frac{2h}{r}\right) \ln\left(\frac{2h}{r}\right)} \text{ F/m}, \quad \text{for } \frac{2h}{r} \gg 1 \qquad (1)$$

$$= \frac{27.8 \, \varepsilon_{r(\text{eff})} \ln\left[1 + \left(\frac{2h}{d}\right)^2\right]}{\left[\ln\left(\frac{2h}{r}\right)\right]^2} \text{ pF/m} \qquad (2a)$$

$$= \frac{0.7 \, \varepsilon_{r(\text{eff})} \ln\left[1 + \left(\frac{2h}{d}\right)^2\right]}{\left[\ln\left(\frac{2h}{r}\right)\right]^2} \text{ pF/in} \qquad (2b)$$

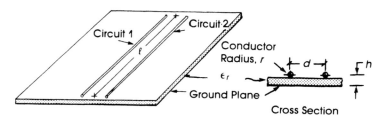

Figure 2.5 Two parallel conductors are spaced a distance d apart and are located h above a ground plane. The circuit returns are via the ground plane.

Example:

Assume that two circuits each contain #28 AWG stranded conductors (effective diameter = 0.015" thus $r = 0.0075"$) which are spaced 1" apart and are parallel for a run of 4". Compute the "stray" capacitance between these conductors without and with a ground plane, assuming $\varepsilon_{r(\text{eff})} = 1$.

From formula set C-1,

$$\frac{C}{l} \approx \frac{0.7\, \varepsilon_{r(\text{eff})}}{\ln\left(\dfrac{d}{r}\right)} \text{ pF/in} \qquad (3)$$

$$C \approx \frac{0.7 \times 1}{\ln\left(\dfrac{1"}{0.0075"}\right)} \times 4" \text{ pF}$$

$$= 0.57 \text{ pF} \qquad (4)$$

(In this case, the two circuits have common grounds, connected by normal size wires. See Fig. 3.3 for a similar geometric configuration.)

Now if a ground plane is located 0.060" below the conductors, forming the common ground for the two circuits, Eq. (2b) yields

$$C \approx \frac{0.7 \times 4" \ln\left[1 + \left(\dfrac{0.120"}{1"}\right)^2\right]}{\ln\left(\dfrac{0.120"}{0.0075"}\right) \ln\left(\dfrac{0.120"}{0.0075"}\right)} \text{ pF}$$

$$= 0.0052 \text{ pF} \qquad (5)$$

Equation (5) shows that the addition of a ground plane has reduced the mutual capacitance by a factor of 109.5.

Derivation:

We will first consider the meaning of mutual capacitance. Analogous to the concept of mutual inductance, discussed in Formula Sets L-3, L-7, L-8, and L-10, the mutual capacitance is defined as the current produced in circuit 2 by the time derivative of the voltage in circuit 1 or:

$$i_2 = C_m \frac{dV_1}{dt} \qquad (6)$$

Thus,

$$C_m = \frac{i_2}{\left(\dfrac{dV_1}{dt}\right)} \qquad (7)$$

where

$\dfrac{dV_1}{dt}$ = the time derivative of V_1

Please note that d, as used in this equation signifies the derivative and not distance. The conductor arrangement shown in Fig. 2.5 is redrawn as an equivalent circuit in Fig. 2.6 to show this concept.

The following derivation is based on the equivalency between the magnetic flux field and the electrostatic potential field. Reference [1] shows that the two fields are indeed identical provided that, for the magnetic field case, the conductors are thin cylinders—not an unreasonable assumption especially at the higher frequencies where the skin effect predominates.

Figure 2.7 illustrates the concept of mutual inductance by using a magnetic field plot. The voltage induced in circuit 2 is related to the current in circuit 1 by the mutual inductance, L_m:

$$V_2 = L_m \frac{di_1}{dt} \qquad (8)$$

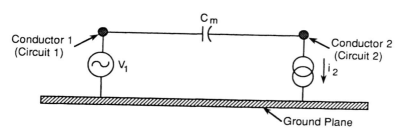

Figure 2.6 The mutual capacitance may be represented by a single capacitor, C_m, connected between circuits 1 and 2. The time-varying voltage on conductor 1 produces a current, i_2, in the circuit 2 load.

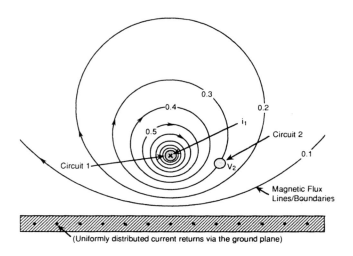

Figure 2.7 Current i_1 flows into conductor 1 setting up a magnetic field containing magnetic potential and continuous magnetic flux lines (tubes). Some of these flux lines link circuit 2. For our purposes, circuit 2 is considered to be "open" and therefore produces no counter flux line since the current is zero. Thus, the field between conductor 2 and the ground plane is due totally to the current in conductor 1.

or

$$L_m = \frac{V_2}{\left(\dfrac{di_1}{dt}\right)} \qquad (9)$$

Mutual inductance can also be defined as the number of flux lines linking circuit 2 caused by a current in circuit 1:

$$L_m = \frac{\Psi_m}{i_1} \text{ H} \qquad (10)$$

Self-inductance is defined as the ratio of total flux lines, Ψ_1, divided by the current, i_1, or

$$L_1 = \frac{\Psi_1}{i_1} \text{ H} \qquad (11)$$

(Strictly speaking, self-inductance is defined as $L_1 = N\Psi_1/i_1$ where N equals the number of turns. In this case, $N = 1$.) Combining Eqs. (10) and (11) to eliminate i_1 yields

$$\frac{L_m}{L_1} = \frac{\Psi_m}{\Psi_1} = \alpha$$

$$= \frac{\text{Flux Lines Common Between Conductors 1 and 2}}{\text{Total Number of Flux Lines Generated by Conductor 1}} \quad (12)$$

For the case shown in Fig. 2.7, the mutual inductance between circuits 1 and 2 is then:

$$L_m \approx 0.3 \, L_1 \quad (13)$$

We will now consider the analogous electric field shown in Fig. 2.8.

In the magnetic field discussed above, it was noted that no counter flux was generated by circuit 2, leaving unchanged the field generated by circuit 1. In this

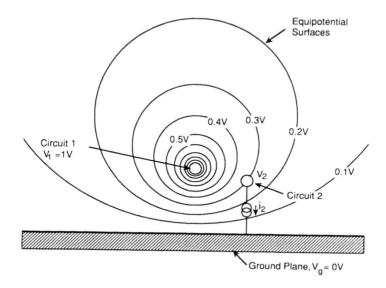

Figure 2.8 Conductor 1, at a potential V_1 volts, produces an electric field consisting of equipotential surfaces and electric flux lines (tubes). In this case, conductor 2 lies on the equipotential line $V_2 = 0.3 \, V$. Current i_2 flows from conductor 2 through a current sink to the ground plane.

case, the electric field remains unchanged if we allow no new charges to be generated on conductor 2. (A land connected to the summing junction of an operational amplifier is a good example. The summing junction is a virtual ground and therefore remains at zero potential. Displacement current resulting from a nearby time-varying field is furnished by the operational amplifier output.)

The current i_2 can be defined as the time rate of change of charge on conductor 2

$$i_2 = \frac{dQ_2}{dt} \tag{14}$$

From the definition of capacitance,

$$Q_2 = C_2 V_2 \tag{15}$$

Differentiating both sides of Eq. (15), we get

$$\frac{dQ_2}{dt} = C_2 \frac{dV_2}{dt} \tag{16}$$

The potential at conductor 2 is a fraction of that of conductor 1:

$$V_2 = \beta V_1 \tag{17}$$

and

$$\frac{dV_2}{dt} = \beta \frac{dV_1}{dt} \tag{18}$$

Equation (7) is

$$C_m = \frac{i_2}{\left(\dfrac{dV_1}{dt}\right)}$$

Combining Eqs. (7) and (14) through (18) we get

$$C_m = \frac{V_2}{V_1} C_2 \tag{19}$$

$$= \beta C_2 \tag{20}$$

Stated a little differently,

$$\frac{C_m}{C_2} = \frac{V_2}{V_1} = \frac{\text{Potential of conductor 2}}{\text{Potential of conductor 1}} = \beta \qquad (21)$$

For the case shown in Fig. 2.8,

$$C_m \approx 0.3 \, C_2 \qquad (22)$$

Since the electric potential field and magnetic flux fields have the same shape for a given geometry, the fields shown in Figs. 2.7 and 2.8 are equivalent. Thus, $\alpha = \beta$ and Eq. (21) equals Eq. (11).

$$\frac{C_m}{C_2} = \frac{L_m}{L_1}$$

or

$$C_m = \frac{L_m C_2}{L_1} \qquad (23)$$

If the conductors are the same radius, r, and the same height, h, above the ground plane, then Eq. (23) becomes

$$C_m = \frac{L_m C_1}{L_1} \qquad (24)$$

Or, on a per unit length basis,

$$\frac{C_m}{l} = \frac{(L_m/l)(C_1/l)}{(L_1 l)} \qquad (25)$$

Formula Sets C-2, L-2, and L-3 give the equations for C_1, L_1, and L_m per unit length, respectively,

$$\frac{C}{l} \approx \frac{2\pi \varepsilon_{r(\text{eff})} \varepsilon_0}{\ln\left(\frac{2h}{r}\right)} \, \text{F/m} \qquad (26)$$

$$\frac{L}{l} \approx \frac{\mu_r \mu_0}{2\pi} \ln\left(\frac{2h}{r}\right) \text{H/m} \qquad (27)$$

$$L_m = \frac{\mu_r \mu_0}{4\pi} \ln\left[1 + \left(\frac{2h}{d}\right)^2\right] \quad (28)$$

Combining Eqs. (25) through (28), we get Eq. (1):

$$\frac{C_m}{l} \approx \frac{\pi \varepsilon_{r(\text{eff})} \varepsilon_0 \ln\left[1 + \left(\frac{2h}{d}\right)^2\right]}{\ln\left(\frac{2h}{r}\right) \ln\left(\frac{2h}{r}\right)} \text{ F/m}$$

Commentary and Conclusions:

1. As noted above, the returns for the circuits are via the ground plane, and the definition of mutual capacitance is based on this premise. If we were to measure directly the capacitance between conductors 1 and 2, we would find that it would be even *greater* in the presence of a ground plane than that calculated from Formula Set C-1 because the two shunt capacitances to ground plane are now in parallel with the main capacitance as shown in Fig. 2.9.
2. If the conductors are not the same radius nor the same height above the ground plane, then Eq. (23) applies.

$$C_m = \frac{L_m C_2}{L_1}$$

Reference [2] gives the mutual capacitance in terms of the mutual inductance, and the line characteristic impedances:

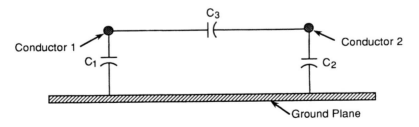

Figure 2.9 The capacitance as measured by an impedance analyzer will be greater than C_3 by a quantity equal to the value of C_1 and C_2 when connected in series. The ground plane is assumed to be floating.

$$C_m = \frac{L_m}{Z_{01}Z_{02}} \text{ F/m} \tag{29}$$

The following equations will show that Eq. (23) is equal to Eq. (29) when the conductors are imbedded in a homogeneous, isotropic and linear medium. Eq. (23) may be restated as

$$C_m = L_m \frac{C_2}{L_1}$$

$$= L_m \sqrt{\frac{C_2}{L_1} \times \frac{C_2}{L_1}} \tag{30}$$

Multiplying the terms within the radical by C_1/C_1 and L_2/L_2, we get

$$C_m = L_m \sqrt{\frac{C_2}{L_1} \times \frac{C_2}{L_1} \times \frac{C_1}{C_1} \times \frac{L_2}{L_2}} \tag{31}$$

Rearranging Eq. (31),

$$C_m = L_m \sqrt{\frac{C_1}{L_1} \times \frac{C_2}{L_2} \times \frac{L_2 C_2}{L_1 C_1}} \tag{32}$$

From Section 1.4, the $LRCZ_0$ analogy,

$$\left(\frac{L}{l} \text{ H/m}\right)\left(\frac{C}{l} \text{ F/m}\right) = \mu\varepsilon \text{ H/m F/m} \tag{33}$$

$$Z_0 = \sqrt{\frac{L}{C}} \; \Omega \tag{34}$$

Thus,

$$Z_{01} = \sqrt{\frac{L_1}{C_1}} \; \Omega \tag{35}$$

$$Z_{02} = \sqrt{\frac{L_2}{C_2}}\,\Omega \qquad (36)$$

and

$$\sqrt{\frac{(L_2/l)(C_2/l)}{(L_1/l)(C_1/l)}} = \sqrt{\frac{\mu\varepsilon}{\mu\varepsilon}} = 1 \qquad (37)$$

Substituting Eqs. (35), (36) and (37) into Eq. (32) yields Eq. (29):

$$C_m = \frac{L_m}{Z_{01}Z_{02}}\,\text{F/m}$$

REFERENCES

1. Boast, William B., *Principles of Electric and Magnetic Fields*, New York, Harper and Brothers, 1956, pp. 205–210, 229, 311.
2. Mohr, R.J., "Coupling between Open Wires over a Ground Plane," *IEEE Symp. on EMC*, July 23–25, 1968, pp. 404–413.

2.2.4 Capacitance between Parallel, Vertical, Flat Conductors, C-4

This formula set introduces the concept of fringing flux. This is an important consideration because fringing increases the value of the capacitance between parallel vertical conductors above that calculated neglecting fringing, often by many times. Please see Formula Set C-6 for the derivation of the fringing factor. An important application is the determination of the capacitance between lands on printed wiring boards such as those shown in Fig. 2.10.

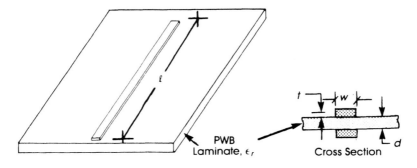

Figure 2.10 Parallel, flat conductors are located on opposite sides of a PWB at a distance d apart.

Equations:

The value of capacitance per unit length is defined as

$$\frac{C}{l} \equiv \varepsilon_r \varepsilon_0 K_{C1}\left(\frac{w}{d}\right) \text{ F/m} \tag{1}$$

$$= 8.84\, \varepsilon_r K_{C1}\left(\frac{w}{d}\right) \text{ pF/m} \tag{2a}$$

$$= 0.225\, \varepsilon_r K_{C1}\left(\frac{w}{d}\right) \text{ pF/in} \tag{2b}$$

where K_{C1} = Capacitive fringing factor (1 or greater). For $d/w \ll 1$, $K_{C1} = 1$.

Fringing Flux:

As previously noted, flux fringing increases the value of the capacitance. Figure 2.11 illustrates this idea with a sketch plotting the approximate flux patterns for a parallel plate capacitor with and without fringing.

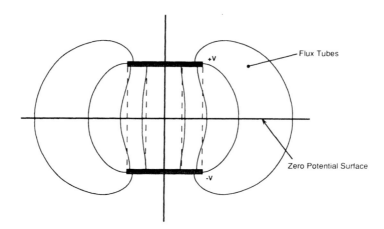

Figure 2.11 The dimensions for this parallel plate capacitor are chosen to show a capacitance increase due to fringing of 2.5 times. Flux lines without fringing are shown dashed. This sketch of the flux pattern shows a total of 10 flux tubes (*versus* 4 in the dashed line case). Thus, the capacitance is 2.5 times greater. (The surrounding medium has a relative dielectric constant, $\varepsilon_r = 1$ and is thus homogeneous for this specific case.)

Example:

Determine the capacitance between two vertical conductors as shown in Fig. 2.10 with these dimensions:

$$t = 0.0028" \text{ (2 oz. copper)} \qquad d = 0.060"$$
$$w = 0.025" \qquad l = 6"$$

Assume that a printed wiring board with epoxy-glass laminate with a relative dielectric constant $\varepsilon_r = 4.5$.

Step 1: Using Fig. 2.12, determine the fringing factor, K_{C1}, for $d/w = 0.060"/0.025" = 2.4$.

Step 2: Determine the capacitance from Eq. (2b) above.

$$C = 0.225 \ \varepsilon_r K_{C1} \left(\frac{w}{d}\right) \times l \text{ pF}$$

$$= 0.225 \times 4.5 \times 2.4 \times \left(\frac{0.025"}{0.060"}\right) \times 6" \text{ pF}$$

$$= 6.1 \text{ pF} \qquad (3)$$

The measured value for this example was 6.4 pF showing excellent correlation.

Derivation:

(Please refer to Formula Set C-6.)

Commentary and Conclusions:

1. This formula set graphically illustrates the effect of fringing on "parallel" plate capacitors. Figure 2.12 indicates that for $d/w = 16$, the actual capacitance is 7.9 times *greater* than would be predicted from direct parallel plate equations, neglecting fringing. This increase could mean the difference between achieving or not meeting the crosstalk requirements.
2. In most PWB designs the land width dimensions, w, are about the same or smaller than the thickness of the circuit boards. For example, for 0.060" thick board and a trace with $w = 0.020"$, $d/w = 3$ (which is not $\ll 1$). Thus, electric flux fringing cannot be ignored.

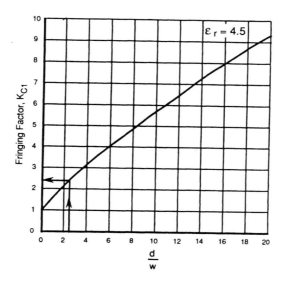

Figure 2.12 The fringing factor, K_{C1}, is determined by entering the graph at the ratio d/w, proceeding upward to the curve and locating K on the left-hand vertical axis as indicated. Thus, for $d/w = 2.4$, $K_{C1} = 2.4$. (This figure was developed in Formula Set C-6.)

2.2.5 Capacitance between Horizontal Flat Conductors, C-5

The capacitance value between horizontal, rectangular conductors is important because this configuration closely approximates conductor lands on printed wiring boards such as those shown in Fig. 2.13.

Equations:

The capacitance per unit length is given approximately by

$$\frac{C}{l} \approx \frac{\pi \varepsilon_{r(\text{eff})} \varepsilon_0}{\ln\left(\dfrac{\pi(d-w)}{w+t} + 1\right)} \text{ F/m} \tag{1}$$

$$= \frac{27.8 \, \varepsilon_{r(\text{eff})}}{\ln\left(\dfrac{\pi(d-w)}{w+t} + 1\right)} \text{ pF/m} \tag{2a}$$

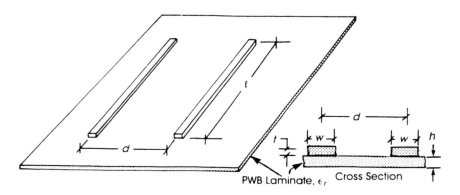

Figure 2.13 Two rectangular conductors are located distance d apart on the top of a circuit board with thickness h and relative dielectric constant, ε_r.

$$= \frac{0.71\, \varepsilon_{r(\text{eff})}}{\ln\left(\dfrac{\pi(d-w)}{w+t}+1\right)} \text{ pF/in} \tag{2b}$$

If $d \gg w$, Eqs. (1), (2a), and (2b) can be further approximated as

$$\frac{C}{l} \approx \frac{\pi \varepsilon_{r(\text{eff})} \varepsilon_0}{\ln\left(\dfrac{\pi d}{w+t}\right)} \text{ F/m} \tag{3}$$

$$= \frac{27.8\, \varepsilon_{r(\text{eff})}}{\ln\left(\dfrac{\pi d}{w+t}\right)} \text{ pF/m} \tag{4a}$$

$$= \frac{0.7\, \varepsilon_{r(\text{eff})}}{\ln\left(\dfrac{\pi d}{w+t}\right)} \text{ pF/in} \tag{4b}$$

For conductors of unequal widths, w_1 and w_2, the capacitance per unit length is given approximately as

$$\frac{C}{l} \approx \frac{2\pi \varepsilon_{r(\text{eff})} \varepsilon_0}{\ln\left[\pi^2 d^2 \left(\dfrac{1}{w_1+t}\right)\left(\dfrac{1}{w_2+t}\right)\right]} \text{ F/m} \tag{5}$$

$$= \frac{55.6\, \varepsilon_{r(\text{eff})}}{\ln\left[\pi^2 d^2 \left(\dfrac{1}{w_1 + t}\right)\left(\dfrac{1}{w_2 + t}\right)\right]} \text{pF/m} \tag{6a}$$

$$= \frac{1.41\, \varepsilon_{r(\text{eff})}}{\ln\left[\pi^2 d^2 \left(\dfrac{1}{w_1 + t}\right)\left(\dfrac{1}{w_2 + t}\right)\right]} \text{pF/in} \tag{6b}$$

If the ratio of the distance, d to the board thickness, h is $\gg 1$, $\varepsilon_{r(\text{eff})} \approx 1$. If $d \approx h$, $\varepsilon_{r(\text{eff})} = (1 + \varepsilon_r)/2$ should be used.

Example 1:

Calculate the capacitance between two horizontal lands, similar to those shown in Fig. 2.13, located on a printed wiring board with a relative dielectric constant $\varepsilon_r = 4.5$. Calculate $\varepsilon_{r(\text{eff})}$ from measured data. The dimensions are

$t = 0.0028''$ (2 oz. copper) $w = 0.025''$
$d = 1''$ $l = 6''$
$h = 0.060''$ $\varepsilon_r = 4.5$

With $d/h = 1''/0.060'' = 16.7 \gg 1$, $\varepsilon_{r(\text{eff})} \approx 1$, the capacitance is

$$C \approx \frac{0.7 \times \varepsilon_{r(\text{eff})} \times 6''}{\ln\left(\dfrac{\pi(1'' - 0.025'')}{0.025'' + 0.0028''} + 1\right)} \text{pF}$$

$$= 0.904\, \varepsilon_{r(\text{eff})}\, \text{pF} \tag{7}$$

Measurements made for these dimensions yielded 0.924 pF. Thus,

$$\varepsilon_{r(\text{eff})} = \frac{0.942}{0.904} = 1.04 \tag{8}$$

Example 2:

Assume that the two traces used in Example 1 are moved closer together so that $d = 0.065''$. Here, $d/h = 0.065''/0.060'' = 1.1$.

$$C \approx \frac{0.7 \times \varepsilon_{r(\text{eff})} \times 6''}{\ln\left[\dfrac{\pi(0.065'' - 0.025'')}{0.025'' + 0.0028''} + 1\right]} \text{ pF}$$

$$= 2.49 \, \varepsilon_{r(\text{eff})} \text{ pF } \textit{versus } 5.41 \text{ pF measured.} \tag{9}$$

The measured effective dielectric constant in this case then, is

$$\varepsilon_{r(\text{eff})} = \frac{5.41}{2.49} = 2.17 \tag{10}$$

Equation Development:

Equation (1) is adapted from Eq. (2), Formula Set C-1, where

$$\frac{C}{l} \approx \frac{\pi \varepsilon_{r(\text{eff})} \varepsilon_0}{\ln\left(\dfrac{d}{r}\right)} \text{ F/m,} \quad \text{for} \quad \frac{2r}{d} \ll 1 \tag{11}$$

By making the perimeter of the round conductor equal to that of the rectangular conductor we will get equal surface areas per unit length for the two geometries:

$$\text{Perimeter} = 2\pi r = 2(w + t) \tag{12}$$

$$r = \frac{w + t}{\pi} \tag{13}$$

Substituting r into Eq. (11) gives

$$\frac{C}{l} = \frac{\pi \varepsilon_{r(\text{eff})} \varepsilon_0}{\ln\left(\dfrac{\pi d}{w + t}\right)} \text{ F/m,} \quad \text{for} \quad \frac{2(w + t)}{\pi d} \ll 1 \tag{14}$$

The limitations of Eq. (14) are illustrated if we let $d = w$. Obviously, the equation is not valid when the conductors are touching and the capacitance per unit length becomes infinite. Adding a correction factor, Δ, to d so that we can get infinite capacitance when $d = w$:

$$\frac{C}{l} = \frac{\pi \varepsilon_{r(\text{eff})} \varepsilon_0}{\ln\left(\dfrac{\pi(d + \Delta)}{w + t}\right)} \text{ F/m,} \quad \text{for} \quad \frac{2(w + t)}{\pi d} \ll 1 \tag{15}$$

For infinite capacitance per unit length when $d = w$:

$$\ln\left(\frac{\pi(w + \Delta)}{w + t}\right) = 0 \tag{16}$$

which means that

$$\left(\frac{\pi(w + \Delta)}{w + t}\right) = 1 \tag{17}$$

Thus Δ is

$$\Delta = \frac{w + t}{\pi} - d \tag{18}$$

Substituting Δ into Eq. (15) yields Eq. (1):

$$\frac{C}{l} \approx \frac{\pi \varepsilon_{r(\text{eff})} \varepsilon_0}{\ln\left(\frac{\pi(d - w)}{w + t} + 1\right)} \text{ F/m}$$

Please see Section 1.3, Electric Field Mapping, for a discussion on round *versus* rectangular conductors.

Equations (5), (6a), and (6b) for conductors of unequal width are derived from preceding Eqs. (11) and (14) as applied to Eq. (4) from Formula Set C-1.

Commentary and Conclusions:

1. EXP C-5A shows excellent correlation between predicted and measured capacitance values.
2. Because the capacitance is governed by the logarithmic term, distance (d) increases do not provide dramatic improvements. If, in Example 1, d is increased from 1" to 2", C decreases to 0.79 pF not by a factor of 2, as might intuitively be expected.
3. Sometimes, the geometric ratio d/h falls between approximately 1 and a number which is much greater than 1. If the application is critical, it is recommended that an effective dielectric constant, $\varepsilon_{r(\text{eff})}$, equal to $(1 + \varepsilon_r)/2$ be used. In some cases the conductors can be fully embedded in dielectric material as in multilayer circuit boards. In this case, the use of an effective dielectric constant, $\varepsilon_{r(\text{eff})} = \varepsilon_r$ should be considered.

2.2.6 Capacitance between a Flat Conductor and a Ground Plane, C-6

Ground planes can be used on circuit boards to provide large crosstalk reductions, as we will see later. This formula set gives the capacitance per unit length between a PWB land and the ground plane, as shown in Fig. 2.14.

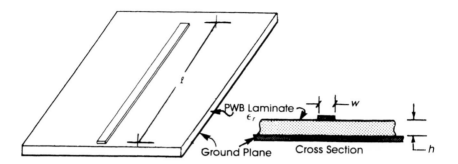

Figure 2.14 A PBW land is located a distance h above a ground plane.

This formula set introduces the relationship between the capacitive fringing factor, K_{C1}, and the inductive fringing factor, K_{L1}. If the medium surrounding the flat conductor and the ground plane is homogeneous, $K_{C1} = K_{L1}$. However, in this case, the flat conductor is separated from the ground plane by the PBW laminate with relative dielectric constant, ε_r. Because the region above the conductor is assumed to have a relative dielectric constant $\varepsilon_r = 1$, the medium surrounding the conductors is not homogeneous.

Note: The capacitance values found here are not to be confused with mutual capacitance.

Equations:

$$\frac{C}{l} = \varepsilon_r \varepsilon_0 K_{C1}\left(\frac{w}{h}\right) \text{F/m} \tag{1}$$

$$= 8.84 \ \varepsilon_r K_{C1}\left(\frac{w}{h}\right) \text{pF/m} \tag{2a}$$

$$= 0.225 \ \varepsilon_r K_{C1}\left(\frac{w}{h}\right) \text{pF/in} \tag{2b}$$

Example:

Assume that a 0.0156" wide land is 11" long and is separated from a ground plane by PWB laminate 0.060" thick with relative dielectric constant $\varepsilon_r = 4.5$. Compare this value with measured data.

Step 1: Determine $2h/w = 2 \times 0.060''/0.0156'' = 7.7$.

Step 2: Enter curve on Fig. 2.15 for $2h/w = 7.7$ and proceed upward to the fringing curve K_{C1}. Find $K_{C1} = 4.8$.

Step 3: Calculate capacitance from Eq. (2b).

$$C = 0.225 \times 4.5 \times 4.8 \times 0.0156''/0.060'' \times 11''$$

$$= 13.9 \text{ pF } versus\ 14.6 \text{ pF measured} \tag{3}$$

Derivation:

The literature indicates that analytical solutions, for this deceptively simple problem, are difficult for the general case, where h is not small compared to w.

Reference [1] gives very accurate approximations for the characteristic impedance, Z_0, and the effective relative dielectric constant, $\varepsilon_{r(\text{eff})}$, as a function of conductor width, w, relative dielectric constant, ε_r, and height, h, above a ground plane. These equations are

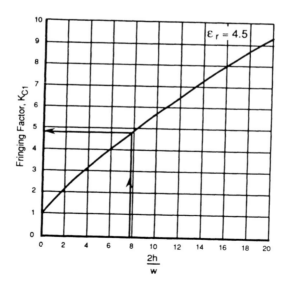

Figure 2.15 In a way similar to Formula Set C-4, the fringing factor, K_{C1}, is determined by entering the graph at the ratio $2h/w$, proceeding upward to the curve and locating K_{C1} on the left-hand vertical axis. Thus, for $2h/w = 7.7$, $K_{C1} = 4.8$.

For $\varepsilon_r = 1$:

$$Z_{0(\varepsilon_r=1)} \approx 60 \ln\left(\frac{8h}{w} + \frac{w}{4h}\right) \Omega, \quad \frac{w}{h} \leq 1 \qquad (4)$$

$$Z_{0(\varepsilon_r=1)} \approx \frac{120\pi}{\frac{w}{h} + 2.42 - 0.44\left(\frac{h}{w}\right) + \left(1 - \frac{h}{w}\right)^6} \Omega, \quad \frac{w}{h} \geq 1 \qquad (5)$$

where

$Z_{0(\varepsilon_r=1)}$ = characteristic impedance with homogeneous medium and $\varepsilon_r = 1$.

For the relative dielectric constant, ε_r:

$$Z_{0(\varepsilon_r)} = \frac{Z_{0(\varepsilon_r=1)}}{\sqrt{\varepsilon_{r(\text{eff})}}} \Omega \qquad (6)$$

where

$Z_{0(\varepsilon_r)}$ = characteristic impedance with a dielectric with relative dielectric constant, ε_r, between the flat conductor and the ground plane;

$\varepsilon_{r(\text{eff})}$ = effective relative dielectric constant:

$$= \frac{\varepsilon_r + 1}{2} + \left(\frac{\varepsilon_r - 1}{2}\right)\left(\frac{1}{\sqrt{1 + \frac{10h}{w}}}\right) \qquad (7)$$

Note: $\varepsilon_{(\text{eff})}$ in Ref. [1] is changed to $\varepsilon_{r(\text{eff})}$ to be consistent with notation used in this book.

From Formula Set Z_0-(ALL), Z_0 is given by

$$Z_0 = \sqrt{\frac{L}{C}} \Omega \qquad (8)$$

$$= \sqrt{\frac{L/l}{C/l}} \Omega \qquad (9)$$

Formula Set L-6 and this formula set define L/l and C/l, respectively, as

$$\frac{L}{l} = \frac{\mu_r\mu_0}{K_{L1}} \left(\frac{h}{w}\right) \text{H/m} \qquad (10)$$

$$\frac{C}{l} = \varepsilon_r\varepsilon_0 K_{C1} \left(\frac{w}{h}\right) \text{F/m} \qquad (11)$$

where

K_{L1} = Inductive fringing factor, dimensionless
K_{C1} = Capacitive fringing factor, dimensionless

Combining Eqs. (9), (10) and (11),

$$Z_0 = \frac{1}{\sqrt{K_{L1}K_{C1}}} \sqrt{\frac{\mu_r\mu_0}{\varepsilon_r\varepsilon_0}} \left(\frac{h}{w}\right) \Omega \qquad (12)$$

With $\mu_r = 1$, $\mu_0 = 4\pi \times 10^{-7}$ H/m and $\varepsilon_0 = 10^{-9}/36\pi$ F/m,

$$Z_{0(\varepsilon_r)} = \frac{120\pi}{\sqrt{K_{L1}K_{C1}}\sqrt{\varepsilon_r}} \left(\frac{h}{w}\right) \Omega \qquad (13)$$

As previously stated, if $\varepsilon_r = 1$, the medium between the conductors is homogeneous and therefore $K_{L1} = K_{C1}$. Solving for K_{L1} by letting $Z_0 = Z_{0(\varepsilon_r=1)}$ in Eq. (13), we get

$$Z_{0(\varepsilon_r=1)} = \frac{120\pi}{K_{L1}\sqrt{1}} \left(\frac{h}{w}\right) \Omega \qquad (14)$$

$$K_{L1} = \frac{120\pi}{Z_{0(\varepsilon_r=1)}} \left(\frac{h}{w}\right) \qquad (15)$$

Note that K_{L1} is dependent only on the relative geometrical dimensions, h and w, and not on the dielectric constant of the material between the flat conductor and the ground plane. K_{C1} is, however, as will be seen in Eq. (16). We can solve for K_{C1} by combining Eqs. (6) and (13) and squaring

Table 2.2
Fringing Factors K_{L1}, K_{C1} and $Z_{0(\varepsilon_r=1)}$, $\varepsilon_{r(\text{eff})}$ versus w/h ($\varepsilon_r = 4.5$)

$\dfrac{w}{h}$	$Z_{0(\varepsilon_r=1)}\Omega$	$\varepsilon_{r(\text{eff})}$	$\dfrac{Z_{0(\varepsilon_r=1)}}{\sqrt{\varepsilon_{r(\text{eff})}}}\Omega$	K_{L1}	K_{C1}	$\dfrac{2h}{w}$
0.100	262.94	2.92	153.76	14.33	9.30	20
0.125	249.56	2.94	145.44	12.08	7.90	16
0.2	221.41	3.00	127.94	8.51	5.68	10
0.25	208.06	3.02	119.66	7.25	4.86	8
0.5	166.82	3.13	94.27	4.52	3.14	4
1.0	126.51	3.28	69.88	2.98	2.17	2
2.5	78.69	3.53	41.87	1.92	1.50	0.8
5.0	49.64	3.76	25.60	1.52	1.27	0.4
10	29.21	3.99	14.62	1.29	1.14	0.2

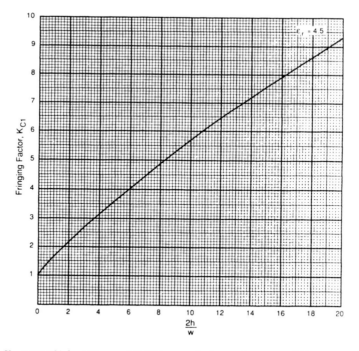

Figure 2.16 K_{C1} versus $2h/w$ with $\varepsilon_r = 4.5$. This curve indicates that, as would be expected, the fringing flux and hence the fringing factor increases with increasing height-to-width ratio, $2h/w$.

$$K_{C1} = \left[\frac{120\pi}{Z_{0(\varepsilon_r=1)}} \sqrt{\frac{\varepsilon_{r(\text{eff})}}{K_{L1}\varepsilon_r}} \left(\frac{h}{w}\right) \right]^2 \tag{16}$$

Table 2.2 lists $Z_{0(\varepsilon_r=1)}$, K_{L1} and K_{C1} for various values of w/h using $\varepsilon_r = 4.5$. K_{C1} for other values of ε_r can be determined in a similar manner.

Figure 2.16 plots the capacitive fringing factor, K_{C1}, versus the geometrical ratio $2h/w$. (K_{L1} is plotted in Formula Set L-6.)

The effective dielectric constant, $\varepsilon_{r(\text{eff})}$ versus $2h/w$ is shown in Fig. 2.17.

In Fig. 2.12, we note that the ratio d/w is used for Formula Set C-4, Vertical Flat Conductors. Figure 2.18 illustrates the reason for this.

Commentary and Conclusions:

1. The solution developed above was easily applied to Formula Set C-4 because the electric fields are equivalent.

REFERENCES

1. Schneider, M.V., "Microstrip Lines for Microwave Integrated Circuits," *Bell System Technical Journal*, Vol. 48, No. 5 (May–June 1969).

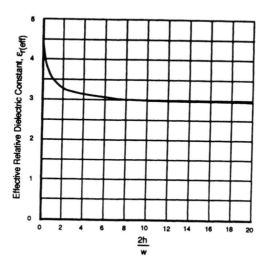

Figure 2.17 $\varepsilon_{r(\text{eff})}$ versus $2h/w$ with $\varepsilon_r = 4.5$. As $2h/w$ approaches zero, $\varepsilon_{r(\text{eff})}$ approaches 4.5, as would be expected because most of the electric flux is totally in the PWB laminate. Conversely, as $2h/w$ becomes large, $\varepsilon_{r(\text{eff})}$ approaches 2.75, the average of the air and laminate dielectric constants.

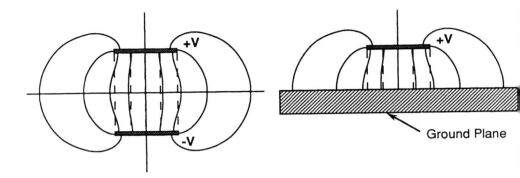

Figure 2.18 The flux pattern for the upper half of two vertical, flat conductors is identical to that for a single flat conductor over a ground plane. Thus, if we make $2h = d$, the correct fringing factor is found in Fig. 2.16.

2.2.7 Mutual Capacitance between Two Horizontal, Flat Conductors Near a Ground Plane, C-7

This is perhaps the most important of the capacitance formula sets because it applies directly to single-sided or multilayer printed wiring boards with a ground plane on one side or as a layer. This formula set shows the dramatic reduction in crosstalk provided by the addition of a ground plane. Figure 2.19 illustrates the geometry.

Equations:

The mutual capacitance/unit length is given approximately by Eq. (1):

$$\frac{C_m}{l} \approx \frac{\varepsilon_r \varepsilon_0}{\pi} K_{L1} K_{C1} \left(\frac{w}{d}\right)^2 \text{ F/m}, \quad \text{for} \quad \frac{2h}{d} < 0.3 \tag{1}$$

$$= 2.81 \, \varepsilon_r K_{L1} K_{C1} \left(\frac{w}{d}\right)^2 \text{ pF/m} \tag{2a}$$

$$= 0.07 \, \varepsilon_r K_{L1} K_{C1} \left(\frac{w}{d}\right)^2 \text{ pF/in} \tag{2b}$$

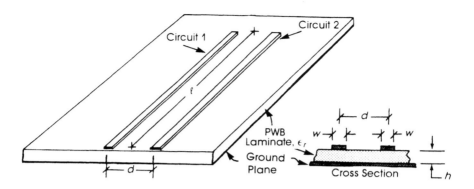

Figure 2.19 Two parallel lands are separated from a ground plane by the circuit board thickness, h.

Example:

Calculate the mutual capacitance between two PWB lands which are 0.025" wide, 1" apart, and 6" long on a 0.060" thick board with $\varepsilon_r = 4.5$.

Step 1: Check the ratio $2h/d$,

$$\frac{2h}{d} = \frac{2 \times 0.060"}{1"} = 0.12 \tag{3}$$

which is less than 0.3. If greater, Eq. (9) should be used.

Step 2: Determine the ratio $2h/d$

$$\frac{2h}{w} = \frac{2 \times 0.060"}{0.025"} = 4.8 \tag{4}$$

Step 3: Determine the fringing factors K_{L1} and K_{C1} from Fig. 2.20.
Step 4: Solving for C_m from Eq. (2b),

$$\frac{C_m}{l} = 0.07 \, \varepsilon_r K_{L1} K_{C1} \left(\frac{w}{d}\right)^2 \text{ pF/in}$$

$$C_m = 0.07 \times 4.5 \times 5.1 \times 3.5 \times \left(\frac{0.025"}{1"}\right)^2 \times 6" \text{ pF}$$

$$= 0.022 \text{ pF} \tag{5}$$

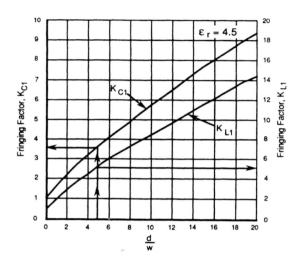

Figure 2.20 To find the fringing factors enter the curve at $2h/w = 4.8$ and find $K_{L1} = 5.1$ and $K_{C1} = 3.5$ on the vertical axes.

In the example of Formula Set C-5 for the same geometry, we had 0.90 pF. The improvement gained by adding the ground plane is a factor of 41.3, which would result in a crosstalk reduction of 32.3 dB. Please see EXP C-5 and EXP C-7 for additional details.

Derivation:

Formula Set C-3 gives the mutual capacitance between two circular conductors of equal height over a ground plane as

$$\frac{C_m}{l} = \frac{\left(\frac{L_m}{l}\right)\left(\frac{C_1}{l}\right)}{\left(\frac{L_1}{l}\right)} \tag{6}$$

where

$\frac{L_m}{l}$ = mutual inductance between conductors over a ground plane per unit length,

$\frac{C_1}{l}$ = capacitance between conductor 1 and the ground plane per unit length,

$\dfrac{L_1}{l}$ = self-inductance of conductor 1 and the ground plane per unit length.

L_m is found in Formula Set L-7 and is approximately

$$\frac{L_m}{l} \approx \frac{\mu_r \mu_0}{4\pi} \ln\left[1 + \left(\frac{2h}{d}\right)^2\right] \text{H/m} \tag{7}$$

Formula Set C-6 gives the capacitance per unit length as

$$\frac{C}{l} = \varepsilon_r \varepsilon_0 K_{C1}\left(\frac{w}{h}\right) \text{F/m} \tag{8}$$

From Formula Set L-6, the inductance per unit length is

$$\frac{L}{l} = \frac{\mu_r \mu_0}{K_{L1}}\left(\frac{h}{w}\right) \text{H/m} \tag{9}$$

Combining Eqs. (6), (7), (8), and (9), we get

$$\frac{C_m}{l} = \frac{\dfrac{\mu_r \mu_0}{4\pi} \ln\left[1 + \left(\dfrac{2h}{d}\right)^2\right] \varepsilon_r \varepsilon_0 K_{C1}\left(\dfrac{w}{h}\right)}{\dfrac{\mu_r \mu_0}{K_{L1}}\left(\dfrac{h}{w}\right)} \text{F/m}$$

$$= \frac{\varepsilon_r \varepsilon_0 K_{L1} K_{C1} \ln\left[1 + \left(\dfrac{2h}{d}\right)^2\right]}{4\pi \left(\dfrac{h}{w}\right)^2} \text{F/m} \tag{10}$$

$$= \frac{\varepsilon_r \varepsilon_0}{4\pi} K_{L1} K_{C1} \left(\frac{w}{h}\right)^2 \ln\left[1 + \left(\frac{2h}{d}\right)^2\right] \text{F/m} \tag{11}$$

Noting that $\ln(1 + x)$ can be expanded,

$$\ln(1 + x) = x - \frac{x^2}{2} + \frac{x^3}{3} - \frac{x^4}{4} + \cdots \tag{12}$$

This series converges for $-1 < x \leq 1$. By neglecting all terms except the first and comparing the value of $\ln(1 + x)$ with x, we find the error to be 14% for $x = 0.3$, 5% for $x = 0.1$, et cetera. We can then simplify Eq. (11) by using Eq. (12) with $x = (2h/d)^2$. This procedure produces Eq. (1):

$$\frac{C_m}{l} \approx \frac{\varepsilon_r \varepsilon_0}{4\pi} K_{L1} K_{C1} \left(\frac{w}{h}\right)^2 \left(\frac{2h}{d}\right)^2$$

$$= \frac{\varepsilon_r \varepsilon_0}{\pi} K_{L1} K_{C1} \left(\frac{w}{d}\right)^2 \text{ F/m} \qquad (13)$$

Commentary and Conclusions:

1. The example shows that a ground plane greatly reduces the mutual capacitance and hence crosstalk between two conductors.
2. As discussed in Formula Set C-3, the return power supply bus land is connected to the ground plane. Thus, the ground plane stays at zero potential with respect to other circuit land and component voltages.
3. Please note that the mutual capacitance is very distance sensitive. Decreasing the spacing by a factor of 10 increases the mutual capacitance by a factor of 100. Inspection of Eq. (1) indicates this is due to the squared term.
4. Subsection 5.2.2, Ground Plane Resistance (R-2) analyzes the "goodness" of the ground plane. The ground plane must be of sufficiently low impedance to prevent significant voltages being introduced between operational amplifier summing junctions and the input signal.
5. The derivation of Eq. (1) is based, in part, on expressions for mutual inductance of circular conductors in which only the height above the ground plane, h, and the distance apart, d, are factors. In a way analogous to Formula Set L-7, the assumption is made that the total electric field potential distribution is approximately the same for flat conductors as it is for circular conductors. Thus, the potential at land 2 due to the voltage on land 1 will be about the same in either case.

2.2.8 Four-Conductor-System Mutual Capacitance, C-8

This formula set has application for determining the mutual capacitance between wiring cable pairs or other four conductor systems. The two circuits are assumed to be both ground, and mutually isolated.

Formula Set C-3 gave the value of the mutual capacitance between two conductors in the presence of a ground plane. This formula set solves for the mutual capacitance between two parallel line sets. Effects of shielding are neglected.

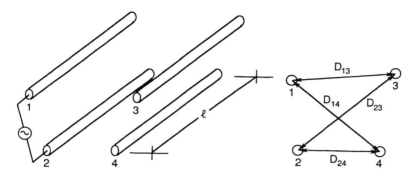

Figure 2.21 Four parallel conductors are spaced at arbitrary distances.

Figure 2.21 illustrates the geometry for two conductor pairs with a voltage impressed between conductors 1 and 2 and voltage pickup appearing on conductors 3 and 4.

Equations:

The mutual capacitance per unit length for the conductors shown in Fig. 2.21 (with conductor radii $r_1 = r_2$ and $r_3 = r_4$) is given for $D_{12} \gg 2r_1$, $D_{34} \gg 2r_3$ approximately as

$$\frac{C_m}{l} \approx \frac{\pi \epsilon_{r(\text{eff})} \epsilon_0 \ln\left(\frac{D_{14} \times D_{23}}{D_{13} \times D_{24}}\right)}{2 \ln\left(\frac{D_{12}}{r_1}\right) \ln\left(\frac{D_{34}}{r_3}\right)} \text{ F/m} \tag{1}$$

$$= 13.9 \, \epsilon_{r(\text{eff})} \frac{\ln\left(\frac{D_{14} \times D_{23}}{D_{13} \times D_{24}}\right)}{\ln\left(\frac{D_{12}}{r_1}\right) \ln\left(\frac{D_{34}}{r_3}\right)} \text{ pF/m} \tag{2a}$$

$$= 0.35 \, \epsilon_{r(\text{eff})} \frac{\ln\left(\frac{D_{14} \times D_{23}}{D_{13} \times D_{24}}\right)}{\ln\left(\frac{D_{12}}{r_1}\right) \ln\left(\frac{D_{34}}{r_3}\right)} \text{ pF/in} \tag{2b}$$

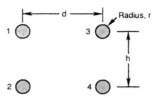

Figure 2.22 The four conductors with equal radii are located on the corners of a rectangle.

Figure 2.22 shows the special case where the conductor arrangement is rectangular and $r_1 = r_2 = r_3 = r_4 = r$.

The distances are

$$D_{14} = D_{23} = \sqrt{d^2 + h^2}$$

$$D_{13} = D_{24} = d$$

For this case, the mutual capacitance per unit length is

$$\frac{C_m}{l} = \frac{\pi \varepsilon_{r(\text{eff})} \varepsilon_0 \ln\left[1 + \left(\frac{h}{d}\right)^2\right]}{\ln\left(\frac{h}{r}\right) \ln\left(\frac{h}{r}\right)} \text{ F/m} \tag{3}$$

$$= \frac{13.9 \, \varepsilon_{r(\text{eff})} \ln\left[1 + \left(\frac{h}{d}\right)^2\right]}{\ln\left(\frac{h}{r}\right) \ln\left(\frac{h}{r}\right)} \text{ pF/m} \tag{4a}$$

$$= \frac{0.35 \, \varepsilon_{r(\text{eff})} \ln\left[1 + \left(\frac{h}{d}\right)^2\right]}{\ln\left(\frac{h}{r}\right) \ln\left(\frac{h}{r}\right)} \text{ pF/in} \tag{4b}$$

Example:

In Formula Set C-3 the mutual capacitance was calculated to be 0.0052 pF for two parallel #28 conductors 1" apart, 4" long, located 0.060" above a ground plane. In this case, we will replace the ground plane with two additional conductors (see Fig. 2.23).

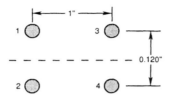

Figure 2.23 For comparison purposes, the circuits are revised by replacing the ground plane with two circular conductors 0.120" below the original.

The mutual capacitance between conductor sets 1-2 and 3-4 is calculated from Eq. (4b) for $\varepsilon_{r(\text{eff})} = 1$ and $r = 0.0075"$.

$$C_m = \frac{0.35 \times 1 \ln\left[1 + \left(\frac{0.120"}{1"}\right)^2\right]}{\ln\left(\frac{0.120"}{0.0075"}\right) \ln\left(\frac{0.120"}{0.0075"}\right)} \times 4" \text{ pF}$$

$$= 0.0026 \text{ pF} \qquad (5)$$

This is exactly one-half the value realized in the Formula Set C-3 example. Figure 2.24 shows the field patterns for the ground plane (Formula Set C-3) and four conductor (Formula Set C-8) configurations.

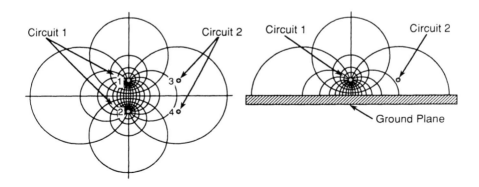

Figure 2.24 The ratio of the voltage appearing on circuit 2's conductors due to circuit 1's potential is the same in both cases. The mutual capacitance for the four conductor case, however, is one-half that for the ground plane configuration because the capacitance between conductors 1 and 2 is reduced by a factor of two. Please see Formula Set C-3 for details.

Derivation:

Equation (23), Formula Set C-3, gives the mutual capacitance as

$$C_m = \frac{L_m C_2}{L_1} \qquad (6)$$

or, on a per unit length basis,

$$\frac{C_m}{l} = \frac{\left(\frac{L_m}{l}\right)\left(\frac{C_2}{l}\right)}{\left(\frac{L_1}{l}\right)} \qquad (7)$$

Formula Set C-1 gives the capacitance per unit length as

$$\frac{C}{l} \approx \frac{\pi \varepsilon_r \varepsilon_0}{\ln\left(\frac{d}{r}\right)} \text{ F/m} \qquad (8)$$

Letting $C = C_2$, $d = D_{34}$, and $r = r_3$, we get

$$\frac{C_2}{l} \approx \frac{\pi \varepsilon_r \varepsilon_0}{\ln\left(\frac{D_{34}}{r_3}\right)} \text{ F/m} \qquad (9)$$

Formula Set L-1 gives the inductance per unit length as

$$\frac{L}{l} \approx \frac{\mu_r \mu_0}{\pi} \ln\left(\frac{d}{r}\right) \text{ H/m} \qquad (10)$$

In a similar way,

$$\frac{L_1}{l} \approx \frac{\mu_r \mu_0}{\pi} \ln\left(\frac{D_{12}}{r_1}\right) \text{ H/m} \qquad (11)$$

Formula Set L-8 gives the mutual inductance per unit length as

$$\frac{L_m}{l} = \frac{\mu_r \mu_0}{2\pi} \ln\left(\frac{D_{14} \times D_{23}}{D_{13} \times D_{24}}\right) \text{H/m} \qquad (12)$$

Combining Eqs. (7), (9), (11), and (12) and letting $\varepsilon_r = \varepsilon_{r(\text{eff})}$, we get Eq. (1):

$$\frac{C_m}{l} \approx \frac{\pi \varepsilon_{r(\text{eff})} \varepsilon_0}{2} \frac{\ln\left(\frac{D_{14} \times D_{23}}{D_{13} \times D_{24}}\right)}{\ln\left(\frac{D_{12}}{r_1}\right) \ln\left(\frac{D_{34}}{r_3}\right)} \text{F/m}$$

Commentary and Conclusions:

1. As noted above, circuits 1-2 and 3-4 must be electrically isolated because common grounds will change the electrostatic field patterns. Further, because the calculated mutual capacitance values can be very small, other circuit components can dominate crosstalk performance in this type of "floating" configuration.

2.2.9 Capacitance between a Flat Conductor and Two Ground Planes (Stripline), C-9

Flat conductors between two ground planes occur in multilayer printed circuit boards and ceramic modules. This formula set, following the methods used in Formula Set C-6, determines the capacitance per unit length between the flat conductor and the two ground planes. Two new fringing factors K_{L2} and K_{C2} are introduced. These account for electrostatic field differences between the microstrip and stripline geometries. The two examples show excellent correlation between values determined by this formula set and those shown in Refs. [1] and [2]. The stripline configuration is shown in Fig. 2.25.

Equations:

The capacitance per unit length is given approximately as follows

$$\frac{C}{l} \approx 2\varepsilon_r \varepsilon_0 K_{C2}\left(\frac{w}{h}\right) \text{F/m} \qquad (1)$$

$$= 17.68 \, \varepsilon_r K_{C2}\left(\frac{w}{h}\right) \text{pF/m} \qquad (2a)$$

$$= 0.45 \, \varepsilon_r K_{C2}\left(\frac{w}{h}\right) \text{pF/in} \qquad (2b)$$

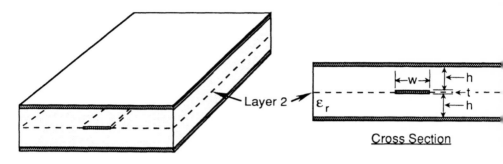

Figure 2.25 A circuit land, located on layer 2, lies between two ground planes, on layers 1 and 3.

Example 1:

Calculate the capacitance per unit length between the land and the ground planes for the configuration shown in Fig. 2.26. Assume that the ground planes are connected. Compare this value with the measured values shown in Ref. [1].

The capacitance per unit length is calculated by using this formula set:

Step 1: Determine $2h/w = (2 \times 0.0075")/0.004" = 3.75$

Step 2: Determine the fringing factor from Fig. 2.27.

Step 3: Calculate the capacitance from Eq. (2b).

$$\frac{C}{l} = 0.45 \times 9.5 \times 3.05 \times (0.004"/0.0075")$$

$$= 6.95 \text{ pF/in } versus \text{ } 6.48 \text{ pF/in shown in Ref. [1]} \quad (3)$$

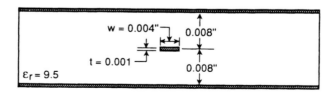

Figure 2.26 A land, embedded in ceramic material with $\varepsilon_r = 9.5$, is located equidistant between two ground planes.

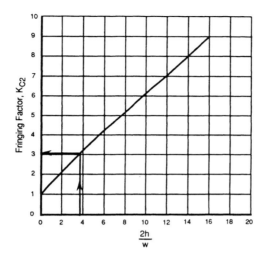

Figure 2.27 For $2h/w = 3.75$, the fringing factor, K_{C2}, = 3.05.

Example 2:

Calculate the capacitance for the ceramic module configuration shown in Fig. 2.28. Compare this value with that determined by finite element analysis modeling discussed in Ref. [2].

The capacitance per unit length is calculated:

Step 1: Determine $2h/w = (2 \times 11.5)/4 = 5.75$.

Step 2: Enter curve on Fig. 2.27 and find $K_{C2} = 4.0$.

Step 3: Applying Eq. (2a), the capacitance to ground for the outer lands is

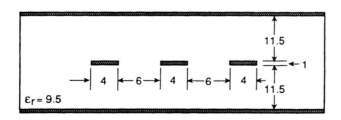

Figure 2.28 Three lands are located between two ground planes.

$$\frac{C}{l} = 17.68 \times 9.5 \times 4.0 \times (4/11.5) \text{ pF/m}$$

$$= 234 \text{ pF/m}$$

$$= 2.34 \text{ pF/cm } \textit{versus } 2.39 \text{ pF/cm found in Ref. [2]}. \qquad (4)$$

Derivation:

As we saw in Formula Set C-6, analytical solutions involving flat conductors are difficult. The approach in this formula set follows the methods used in Formula Set C-6 by determining the fringing factors K_{L2} and K_{C2} from the stripline characteristic impedance. Reference [3] determines the characteristic impedance with the use of resistive paper. The flat conductor and the ground planes are drawn on the resistive paper with a highly conductive paint. The resistance between the flat conductor and the ground planes is measured for various conductor widths. The characteristic impedance is then determined by methods similar to those described in Section 1.4, the $LRCZ_0$ Analogy. Please note that the medium between the conductors is homogeneous. Figure 2.29 shows the results of the resistance measurements converted into Z_0.

Paralleling the derivation used in Formula Set C-6, we have from Formula Set Z_0-(ALL):

$$Z_0 = \sqrt{\frac{L}{C}} \, \Omega \qquad (5)$$

$$= \sqrt{\frac{L/l}{C/l}} \, \Omega \qquad (6)$$

Formula Set L-9 and this formula set define L/l and C/l respectively, as

$$\frac{L}{l} = \frac{\mu_r \mu_0}{2 K_{L2}} \left(\frac{h}{w}\right) \text{H/m} \qquad (7)$$

$$\frac{C}{l} = 2 \varepsilon_r \varepsilon_0 K_{C2} \left(\frac{w}{h}\right) \text{F/m} \qquad (8)$$

where

K_{L2} = inductive fringing factor, dimensionless;
K_{C2} = capacitance fringing factor, dimensionless.

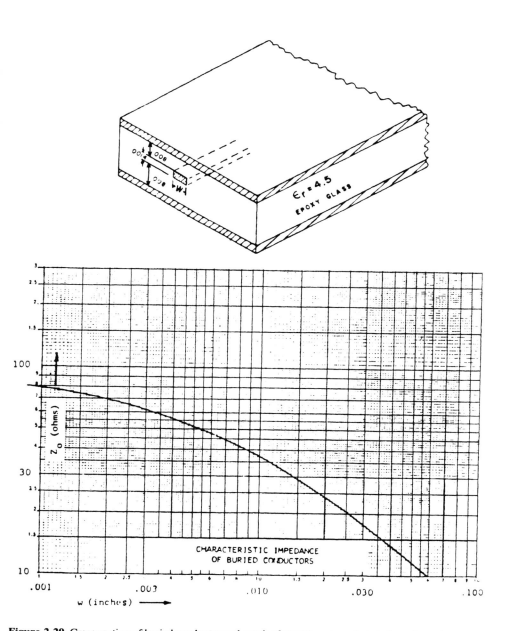

Figure 2.29 Cross section of buried conductor and graph of resulting characteristic impedance (reprinted with permission from Ref. [3], p. 751 © 1967 IEEE).

Combining Eqs. (6), (7), and (8),

$$Z_0 = \frac{1}{2\sqrt{K_{L2}K_{C2}}} \sqrt{\frac{\mu_r\mu_0}{\varepsilon_r\varepsilon_0}} \left(\frac{h}{w}\right) \Omega \qquad (9)$$

With $\mu_r = 1$, $\mu_0 = 4\pi \times 10^{-7}$ H/m, and $\varepsilon_0 = 10^{-9}/36\pi$ F/m,

$$Z_0 = \frac{60\pi}{\sqrt{K_{L2}K_{C2}}\sqrt{\varepsilon_r}} \left(\frac{h}{w}\right) \Omega \qquad (10)$$

Because the medium between the conductors can be considered to be homogeneous, $K_{L2} = K_{C2}$, and therefore

$$Z_{0(\varepsilon_r)} = \frac{60\pi}{K_{C2}\sqrt{\varepsilon_r}} \left(\frac{h}{w}\right) \Omega \qquad (11)$$

$$K_{C2} = \frac{60\pi}{Z_{0(\varepsilon_r)}\sqrt{\varepsilon_r}} \left(\frac{h}{w}\right) \qquad (12)$$

The Z_0 values shown in Fig. 2.29 are $Z_{0(\varepsilon_r)}$ because the surrounding medium uses $\varepsilon_r = 4.5$. Table 2.3 lists the values of $Z_{0(\varepsilon_r)}$ from Fig. 2.29.

The fringing factors listed in Table 2.3 are plotted in Fig. 2.30 as a function of $2h/w$.

Table 2.3
Fringing Factors $K_{L2} = K_{C2}$ versus Conductor Width, $w(\varepsilon_r = 4.5)$

w''	$Z_{0(\varepsilon r)} \Omega$	$K_{L2} = K_{C2}$	$\dfrac{2h}{w}$
0.001	79	9.0	16
0.00114	78	8.0	14
0.00133	76	7.0	12
0.0016	73	6.1	10
0.002	69.5	5.1	8
0.00266	64	4.2	6
0.004	56	3.2	4
0.008	42	2.1	2
0.016	27.8	1.6	1
0.032	16.7	1.3	0.5
0.053	11	1.2	0.3

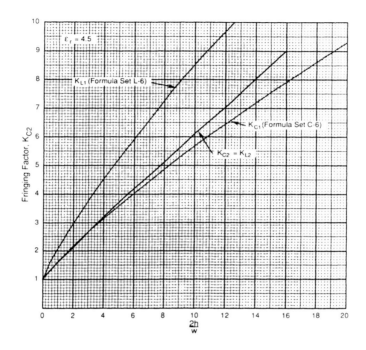

Figure 2.30 $K_{C2} = K_{L2}$ versus $2h/w$. This curve shows that the fringing factor K_{C2} is slightly higher than K_{C1} shown in Formula Set C-6 but K_{L2} is somewhat lower than K_{L1} plotted in Formula Set L-6.

Reference [4] provides general formulas and curves for stripline geometries. This reference determines the quantity $\sqrt{\varepsilon_r}\, Z_0$ as a function of w/b with t/b as a parameter, where w = flat conductor width, t = thickness, and b = distance between the ground planes.

For the geometry shown in Fig. 2.29, $t/b = 0.0014''/0.0174'' = 0.08$. Table 2.4 compares $Z_{0(\varepsilon_r)}$ from Ref. [3] with $Z'_{0(\varepsilon_r)}$ developed from curves shown in Ref. [4]. (The notation $Z'_{0(\varepsilon_r)}$ means that the values were derived from Ref. [4].)

Table 2.4 shows a very good correlation between the two characteristic impedance values obtained by two independent methods. For other t/b values, Reference [4] should be consulted. Table 2.5 indicates $\sqrt{\varepsilon_r}\, Z'_0$ value ranges from this reference. Reference [4] also calculates $\sqrt{\varepsilon_r}\, Z'_0$ values for cases where the flat conductor is not equidistant between the two ground planes.

Commentary and Conclusions:

1. The solution for this configuration is obtained by applying the methods used in Formula Set C-6. Here we can consider that C-10 is made up of two capacitors,

Table 2.4
$Z_{0(\varepsilon_r)}$ versus $Z'_{0(\varepsilon_r)}$ ($t = 0.0014''$, $b = 0.0174''$, and $t/b = 0.08$)

w/b	w''	$Z_{0(\varepsilon_r)}\Omega$	$Z'_{0(\varepsilon_r)}\Omega$
3.0	0.052	11.2	11.2
1.0	0.0174	26	27.5
0.5	0.0087	40	40.7
0.1	0.00174	72	70.5

Table 2.5
$\sqrt{\varepsilon_r} Z'_0$ versus w/b and t/b from Ref. [4]

w/b	$\sqrt{\varepsilon_r} Z'_0 \Omega (t/b = 0)$	$\sqrt{\varepsilon_r} Z'_0 \Omega (t/b = 0.25)$
0.1	193	110
1.0	67	45
4.0	22	19

one from the conductor to the bottom ground plane, the other to the top. These are in parallel, because the ground planes are connected together. In essence, the capacitance values determined by Formula Set C-6 are multiplied by a factor of two and a new fringing factor K_{C2} is used to account for the electrostatic field differences.

2. We have thus seen that this relatively easy method of determining capacitance enjoys very good correlation with the values shown in Refs. [1] and [2].

REFERENCES

1. "Kyocera Design Guidelines Multilayer Ceramic," CAT/1T8205TDN, San Diego, California, Kyocera International, p. 9.
2. Olsen, L.T., "Application of the Finite Element Method to Determine the Electrical Resistance, Inductance, Capacitance Parameters for the Circuit Package Environment," *Trans. IEEE*, Vol. CHMT-5, No. 4, December 1982, pp. 486–492.

3. Catt, I., "Crosstalk (Noise) in Digital Systems," *Trans. IEEE*, Vol. EC-16, No. 6, December 1967, pp. 743–763.
4. Howe, Harlan, *Stripline Circuit Design*, Norwood, MA, Artech House, 1974, pp. 33–40.

2.2.10 Mutual Capacitance between Two Flat Conductors Near Two Ground Planes, C-10

The value of the mutual capacitance per unit length between two long, flat conductors near two ground planes, as shown in Fig. 2.31, is obtained by using the methods of Formula Set C-7. Interestingly enough, the results are the same if K_{C1} is replaced with K_{C2}. This is because the electric field created by the first conductor produces almost the same potential on the second conductor in both the single and double ground plane cases. The slight difference results from the value of the fringing factor.

The formulas give values which provide very good correlation with results shown in Refs. [1] and [2].

Equations:

The mutual capacitance per unit length is

$$\frac{C_m}{l} = \frac{\varepsilon_r \varepsilon_0}{\pi} (K_{C2})^2 \left(\frac{w}{d}\right)^2 \text{ F/m} \tag{1}$$

$$= 2.81 \, \varepsilon_r (K_{C2})^2 \left(\frac{w}{d}\right)^2 \text{ pF/m} \tag{2a}$$

$$= 0.07 \, \varepsilon_r (K_{C2})^2 \left(\frac{w}{d}\right)^2 \text{ pF/in} \tag{2b}$$

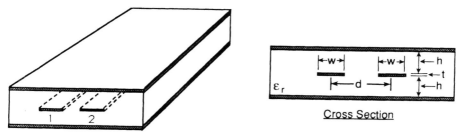

Cross Section

Figure 2.31 Conductors 1 and 2 are embedded between two ground planes. Conductor 1 and the ground planes form circuit 1, while conductor 2 and the ground planes form circuit 2. The returns for each circuit are via the ground planes.

Example 1:

Determine the mutual capacitance between two lands in a ceramic module using two ground planes. Both ground planes are connected to circuit ground forming the returns for the two circuits. Compare the results with values measured in Ref. [1].

The mutual capacitance per unit length is now calculated:

Step 1: From Formula Set C-9, $K_{C2} = 3.2$, and from Fig. 2.32, $d = 0.020''$.

Step 2: Calculate the mutual capacitance from Eq. (2b):

$$\frac{C}{l} = 0.07 \times 9.5 \times (3.2)^2 \times (0.004''/0.020'')^2$$

$$= 0.27 \text{ pF/in } versus \text{ } 0.28 \text{ pF/in shown in Ref. [1]}. \quad (3)$$

Example 2:

Determine the mutual capacitance between lands 1 and 2 in the ceramic module shown in Fig. 2.33. Compare results with those listed in Ref. [2].

The mutual capacitance per unit length is calculated as above:

Step 1: From Formula Set C-9, Example 2, $K_{C2} = 4.0$, and from Fig. 2.33 above, $w = 4$ and $d = 10$ units, respectively.

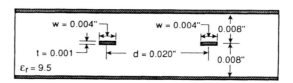

Figure 2.32 In this ceramic module, two 0.004" lands are spaced 0.020" apart and are 0.008" from the ground planes. The relative dielectric constant, ε_r, is 9.5.

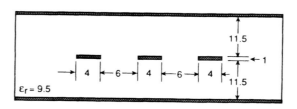

Figure 2.33 Three lands are located between two ground planes in a ceramic module with $\varepsilon_r = 9.5$. Reference [2] calculates the mutual capacitance between land sets 1–2, 2–3, and 1–3 by finite element analysis methods.

Step 2: Calculate the mutual capacitance between lands 1 and 2 from Eq. (2a):

$$\frac{C_m}{l} = 2.81 \times 9.5 \times (4.0)^2 \times (4/10)^2$$

$$= 68.4 \text{ pF/m}$$

$$= 0.68 \text{ pF/cm } versus \text{ 0.58 pF/cm shown in Ref. [2].} \quad (4)$$

Commentary and Conclusions:

1. As noted in the summary, the mutual capacitance formulas for C-7 and this formula set are identical. A slight difference actually exists because a small number of the total flux lines travel in the air above the board in the single-ground plane case. In the double ground-plane case, ideally, all of the electric flux is contained between the ground planes.
2. The two examples demonstrate that the values obtained by this formula set correlate well with data from Refs. [1] and [2].

REFERENCES

1. "Kyocera Design Guidelines Multilayer Ceramic," CAT/1T8205TDN, p.9, Kyocera International, San Diego, California.
2. Olsen, L.T., "Application of the Finite Element Method to Determine the Electrical Resistance, Inductance, Capacitance Parameters for the Circuit Package Environment," *Trans. IEEE*, Vol. CHMT-5, No. 4, December 1982, pp. 486–492.

2.2.11 Capacitance of Coaxial Cables, C-11

Coaxial cables are frequently used for transmitting both high and low level signals between electronic assemblies. Theoretically, these cables produce no electric fields outside the cable, and conversely are immune from uniform external fields. Practical limitations, of course, prevent complete achievement of this ideal situation.

This formula set compares calculated capacitance values with those given for some commercially available cables, as seen in Table 2.6. Figure 2.34 illustrates a coaxial cable and shows the dimensions of interest.

Equations:

Reference [1] gives the capacitance per unit length of a long coaxial cable as

$$\frac{C}{l} = \frac{2\pi\varepsilon_r\varepsilon_0}{\ln\left(\frac{r_2}{r_1}\right)} \text{ F/m} \quad (1)$$

Table 2.6
Calculated *versus* Listed Values for Coaxial Cables

Type	Z_0 (Ω)	ε_r^*	$2r_2$	$2r_1$	Calc. (pF/ft)	Listed (pF/ft)
RG-6/U	75	1.64	0.180"	0.037"	17.5	17.3
RG-6A/U	75	2.3	0.185"	0.028"	20.6	20.5
9393	93	1.64	0.064"	0.010"	14.9	14.0
9252	50	2.3	0.096"	0.028"	31.6	30.8

*ε_r = 1.64 for cellular polyethylene, 2.3 for solid polyethylene.
Source: "Master Catalog," Belden Wire and Cable, P.O. 1989, Box 1980, Richmond, IN, 47375, pp 118–143 and p. 354.

$$= \frac{55.6\, \varepsilon_r}{\ln\left(\dfrac{r_2}{r_1}\right)} \text{ pF/m} \qquad (2a)$$

$$= \frac{1.41\, \varepsilon_r}{\ln\left(\dfrac{r_2}{r_1}\right)} \text{ pF/in} \qquad (2b)$$

Example:

Calculate the capacitance per foot for several commercially available cables and compare these values with those listed in Ref. [2].

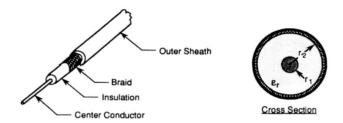

Figure 2.34 Coaxial cables are typically constructed with a solid center conductor surrounded by dielectric insulation, a braided or foil return conductor and an outer insulation sheath.

Equation (2b) is used to calculate the values. Type RG-6/U has $r_1 = 0.0185''$ and $r_2 = 0.090''$:

$$C = \frac{1.41 \times 1.64}{\ln\left(\dfrac{0.090''}{0.0185''}\right)} \times 12'' \text{ pF}$$

$$= 17.5 \text{ pF/ft (for type RG-6/U)} \tag{3}$$

REFERENCES

1. Johnk, C.T.A., *Engineering Electromagnetic Fields and Waves*, New York, John Wiley and Sons, 1973, p. 205.
2. Master Catalog," Belden Wire and Cable, Richmond, IN, 1989, pp. 118–143 and p. 354.

2.2.12 Capacitance between Two Small Spheres, C-12

The expression for capacitance between two small spheres can be useful for estimating the capacitance between electronic components which approximate spherical shapes. Examples are turret terminals, dot contacts, *et cetera*. See Fig. 2.35.

Equations:

Reference [1] gives the capacitance for $a_1, a_2 \ll 2d$ as

$$C \approx 4\pi\varepsilon_r\varepsilon_0 \frac{1}{\left(\dfrac{a_1 + a_2}{a_1 a_2} - \dfrac{1}{d}\right)} \text{ F (dimensions in meters)} \tag{1)*}$$

$$= 111.1 \; \varepsilon_r \frac{1}{\left(\dfrac{a_1 + a_2}{a_1 a_2} - \dfrac{1}{d}\right)} \text{ pF (dimensions in meters)} \tag{2a}$$

$$= 2.82 \; \varepsilon_r \frac{1}{\left(\dfrac{a_1 + a_2}{a_1 a_2} - \dfrac{1}{d}\right)} \text{ pF (dimensions in inches)} \tag{2b}$$

---

*Note: 2c in Ref. [1] is changed to 2d to be consistent with notation used in this book.

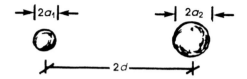

Figure 2.35 Two spheres of unequal radii, a_1 and a_2, are spaced by a distance of $2d$ apart.

Example:

Calculate the capacitance between two small spheres, each 0.010″ in diameter, with $2d = 1″$. Compare this value with that calculated between two circular 0.010″ diameter conductors, 0.010″ long computed from Formula Set C-1. Assume that $\varepsilon_r = 1.0$. The latter will give a rough order of magnitude check when using Formula Set C-1 for very short lines.

From Eq. (2b),

$$C \approx 2.82\, \varepsilon_r \frac{1}{\left(\dfrac{a_1 + a_2}{a_1 a_2} - \dfrac{1}{d}\right)} \text{ pF}$$

$$= 2.82 \times 1 \times \frac{1}{\left(\dfrac{0.005″ + 0.005″}{(0.005″)^2} - \dfrac{1}{0.5″}\right)} \text{ pF}$$

$$= 0.0071 \text{ pF} \tag{3}$$

For the two cylinders, using Formula Set C-1,

$$C \approx \frac{0.7\, \varepsilon_r}{\ln\left(\dfrac{d}{r}\right)} \times l \text{ pF}$$

$$= \frac{0.7 \times 1}{\ln\left(\dfrac{1″}{0.005″}\right)} \times 0.010″ \text{ pF} \tag{4}$$

$$= 0.0013 \text{ pF} \tag{5}$$

Commentary and Conclusions:

1. We see that, in this rather extreme case, the predicted capacitance value for two very short cylinders, when using the formula for long, parallel, circular conductors, will be low by a factor of about 5.5. This is due primarily to the end fringing flux, neglected in Eq. (4).

REFERENCES

1. Weber, Ernst, *Electromagnetic Fields—Theory and Applications*, New York, John Wiley and Sons, 1950, p. 98.

2.3 INDUCTANCE

2.3.1 Self-Inductance of Two Circular Conductors, L-1

The self-inductance value for two long, parallel, circular conductors (as seen in Fig. 2.36) can be used to determine the series inductance of filter capacitor leads, the reactance of twisted-pair ac power lines, the inductance in series with *pulsewidth modulator* (PWM) switching transistors, and many other applications.

Equations:

The external self-inductance per unit length for two conductors of equal radii is given by Ref. [1] as

$$\frac{L}{l} \approx \frac{\mu_r \mu_0}{\pi} \ln\left(\frac{d}{r}\right) \text{H/m} \tag{1}$$

For air (or vacuum), where $\mu_r \approx 1$,

$$\frac{L}{l} \approx 0.4 \ln\left(\frac{d}{r}\right) \mu\text{H/m} \tag{2a}$$

$$= 0.01 \ln\left(\frac{d}{r}\right) \mu\text{H/in} \tag{2b}$$

Example:

Calculate the self-inductance per foot for conductors 1–2 and for 1–25 for the flat cable described in Formula Set C-1 and shown in Fig. 2.37. Compare these values with measurements made at $f = 100$ kHz.

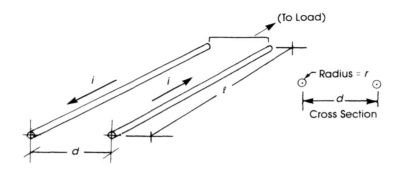

Figure 2.36 Two long circular conductors each have radius r and are separated by distance d. Current i flows down one conductor to the load and back the other.

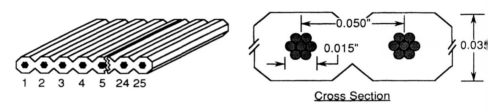

Figure 2.37 This flat cable consists of 25 parallel #28 stranded conductors on 0.050" centers with polyvinyl chloride (PVC) insulation. The effective radius of each conductor is $r \approx 0.0075"$.

Using Eq. (2b) we get for conductor pairs 1–2:

$$\frac{L}{l} \approx 0.01 \ln\left(\frac{d}{r}\right) \mu\text{H/in}$$

$$L = 0.01 \ln\left(\frac{0.050"}{0.0075"}\right) \times 12" \mu\text{H}$$

$$= 0.23 \ \mu\text{H} \ versus \ 0.28 \ \mu\text{H measured} \quad (3)$$

For pairs 1–25,

$$L = 0.01 \ln\left(\frac{24 \times 0.050"}{0.0075"}\right) \times 12" \ \mu\text{H}$$

$$= 0.61 \ \mu\text{H} \ versus \ 0.65 \ \mu\text{H measured} \quad (4)$$

For $f = 100$ kHz, the depth of penetration in copper is 0.008". Thus, the current flowing at the centers of the conductors is related as $1/e$ to that at the surface. The inductance is primarily due to the flux external to the conductor.

Commentary:

1. Reference [1] gives the low-frequency self-inductance for parallel conductors of radii a_1 and a_2, assuming uniform current density within the conductors, as

$$L = \frac{\mu_0 l}{4\pi} + \frac{\mu_0 l}{2\pi} \ln\left(\frac{d^2}{a_1 a_2}\right) \text{H/m} \tag{5}$$

or

$$\frac{L}{l} = \frac{\mu_0}{4\pi} + \frac{\mu_0}{2\pi} \ln\left(\frac{d^2}{a_1 a_2}\right) \text{H/m} \tag{6}$$

Letting $a_1 = a_2 = r$, we get

$$\frac{L}{l} = \frac{\mu_0}{4\pi} + \frac{\mu_0}{2\pi} \ln\left(\frac{d^2}{r^2}\right) \text{H/m}$$

$$= \frac{\mu_0}{4\pi} + \frac{\mu_0}{\pi} \ln\left(\frac{d}{r}\right) \text{H/m} \tag{7}$$

$$= \left[0.1 + 0.4 \ln\left(\frac{d}{r}\right)\right] \times 10^{-6} \text{ H/m} \tag{8}$$

$$= \left[0.00254 + 0.010 \ln\left(\frac{d}{r}\right)\right] \mu\text{H/in} \tag{9}$$

The first term of Eq. (6), $\mu_0/4\pi = 0.1$ μH/m $= 0.00254$ μH/in, represents the internal inductance due to the magnetic field within the conductors, and is independent of conductor radii and spacing. In the example, $d/r = 0.050"/0.0075" = 6.67$ and thus $0.010 \ln 6.67 = 0.019$ μH/in. In low frequency applications, neglecting the first term results in a $0.00254/0.019 = 0.13 = 13\%$ error. Since at high frequencies, most of the current flows on the conductor outer surface, the first term must be neglected. If this is done, Eqs. (8) and (9) become the same as Eqs. (2a) and (2b).

REFERENCES

1. Johnk, C.T.A., *Engineering Electromagnetic Fields and Waves*, John Wiley and Sons, 1973, p. 322.

2.3.2 Self-Inductance of a Circular Conductor and a Ground Plane, L-2

The value of the self-inductance per unit length of a long circular conductor and a conducting ground plane, as shown in Fig. 2.38 is used in the derivation of the mutual inductance between two parallel conductors, and for other applications.

Equations:

The external self-inductance per unit length is half that given in Formula Set L-1, and is

$$\frac{L}{l} \approx \frac{\mu_r \mu_0}{2\pi} \ln\left(\frac{2h}{r}\right) \text{H/m} \tag{1}$$

For air (or vacuum), where $\mu_r = 1$,

$$\frac{L}{l} \approx 0.2 \ln\left(\frac{2h}{r}\right) \mu\text{H/m} \tag{2a}$$

$$= 0.005 \ln\left(\frac{2h}{r}\right) \mu\text{H/in} \tag{2b}$$

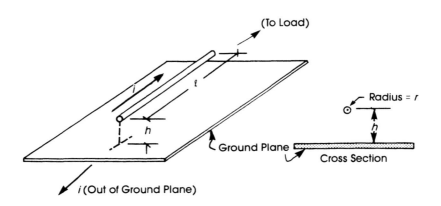

Figure 2.38 A long circular conductor spaced distance h over a ground plane carries a current i. The return current is via the ground plane.

Commentary and Conclusions:

1. The inductance calculated in Eq. 1 is exactly half that for two parallel conductors if $d = 2h$. This is because only half the number of flux lines are generated for the same amount of current.
2. The remarks, approximations and variations appearing in Formula Set L-1 apply equally here.

2.3.3 Mutual Inductance between Two Conductors Near a Ground Plane, L-3

The value of the mutual inductance between two parallel circular conductors near a ground plane can be used for the calculation of inductive coupling (crosstalk) between circuits. An example: wires running close to a metal chassis where the return currents are via the chassis. (For flat conductors, please see Formula Set L-7.)

Equations:

The mutual inductance per unit length is given by

$$\frac{L_m}{l} = \frac{\mu_r \mu_0}{4\pi} \ln\left[1 + \left(\frac{2h}{d}\right)^2\right] \text{H/m} \tag{1}$$

For air (or vacuum) where $\mu_r \approx 1$,

$$\frac{L_m}{l} = 0.1 \ln\left[1 + \left(\frac{2h}{d}\right)^2\right] \mu\text{H/m} \tag{2a}$$

$$= 0.00254 \ln\left[1 + \left(\frac{2h}{d}\right)^2\right] \mu\text{H/in} \tag{2b}$$

Example:

Calculate the mutual inductance between 2 parallel #22 wires 4" long with $d = 1"$ located $h = 0.060"$ above a ground plane. Using Eq. (2b),

$$\frac{L_m}{l} = 0.00254 \ln\left[1 + \left(\frac{2h}{d}\right)^2\right] \mu H/in$$

$$L_m = 0.00254 \ln\left[1 + \left(\frac{2 \times 0.060''}{1''}\right)^2\right] \times 4'' \mu H$$

$$= 0.000145 \, \mu H$$

$$= 0.145 \, nH \qquad (3)$$

Referring to Fig. 2.39, calculate the voltage induced in loop 2 as a result of 10 mA rms flowing in loop 1 at a frequency of 50 kHz. From the definition of mutual inductance, the voltage induced in loop 2 is

$$V_2 = L_m \frac{di_1}{dt} \qquad (4)$$

The current flowing in loop 1 is

$$i_1 = 0.010 \sqrt{2} \sin \omega t \, A \qquad (5)$$

and

$$\frac{di_1}{dt} = 0.010 \times \sqrt{2} \, \omega \cos \omega t = 0.010 \times 2\pi f \cos \omega t \qquad (6)$$

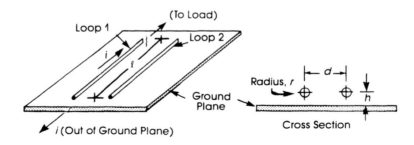

Figure 2.39 A time-varying current, i, flows down the conductor in loop 1 to the load and returns via the ground plane. This current induces a voltage in loop 2 which is proportional to the mutual inductance between the two loops.

Thus, the magnitude of the induced voltage in loop 2 is

$$|V_2| = L_m \left|\frac{di_1}{dt}\right|$$

$$= 0.145 \times 10^{-3} \times 10^{-2} \sqrt{2} \times 2\pi \times 5 \times 10^4 \ \mu V \qquad (7)$$

$$= 0.64 \ \mu V \text{ peak}$$

$$= 0.46 \ \mu V \text{ rms} \qquad (8)$$

$$= -127 \text{ dB}/1 \text{ V}$$

Commentary and Conclusions:

1. If the ground plane were replaced by conductors located h below the plane, we would have from Formula Set L-8, Example 2:

$$\frac{L_m}{l} = 0.005 \ln\left[1 + \left(\frac{h'}{d}\right)^2\right] \mu H/\text{in} \qquad (9)$$

Equation (9) is derived in Formula Set L-8 from Ref. [1], which gives the mutual inductance between conductor pairs 1–2 and 3–4 as

$$\frac{L_m}{l} = \frac{\mu}{2\pi} \ln\left(\frac{D_{14} \times D_{23}}{D_{13} \times D_{24}}\right) H/m \qquad (10)$$

Because $h' = 2h$ in this case, Eq. (9) gives exactly twice the mutual inductance as Eq. (2b). This is the same ratio that we obtained between Formula Set L-1 and L-2 for the reasons stated in L-2; that is, twice as many magnetic flux lines are generated in L-1 as are in L-2.

2. Reference [1] notes that the mutual inductance is independent of conductor size or whether they are hollow or solid as long as they are circular and the current density is uniform. Therefore, neither Eq. (1) nor Eq. (9) contain conductor dimensions. Please see Formula Set L-8 for more information on conductor mutual inductance.

3. The ground plane reduces the mutual inductance between conductors by a factor of 2 (6 dB). If, in the example, we replaced the ground plane with two #22 conductors each 0.060″ directly beneath the existing conductors, the mutual inductance would be calculated from Comment 1 as follows

$$\frac{L_m}{l} = 0.005 \ln\left[1 + \left(\frac{2h}{d}\right)^2\right] \mu\text{H/in} \quad (11)$$

$$L_m = 0.005 \ln\left[1 + \left(\frac{2 \times 0.060''}{1''}\right)^2\right] \times 4'' \, \mu\text{H}$$

$$= 0.00029 \, \mu\text{H}$$

$$= 0.29 \text{ nH} \quad (12)$$

4. The value of mutual inductance is useful for calculating crosstalk levels where the coupling is primarily magnetic. For higher circuit impedance levels, the mutual capacitance between the conductors can become dominant. Crosstalk Analysis CTL-1 discusses this in detail.
5. Reference [2] presents a thorough analysis of crosstalk between wires above a ground plane. The author derives equations, calculates expected values for both inductive and capacitive crosstalk, and supports the findings with experimental data.

REFERENCES

1. Rogers, W.E., *Introduction to Electric Fields*, New York, McGraw-Hill, 1950, p. 316.
2. Mohr, R.J., "Coupling between Open Wires over a Ground Plane," *IEEE Symp. EMC*, July 23–25, 1968, pp. 404–413.

2.3.4 Self-Inductance of Vertical Flat Conductors, L-4

The self-inductance of a loop consisting of long parallel, vertical flat conductors forms an impedance. Current flowing through this impedance produces a voltage drop contributing to ground and power supply bus noise (see Fig. 2.40).

Equations:

The self-inductance per unit length is defined as

$$\frac{L}{l} \approx \frac{\mu_r \mu_0}{K_{L1}} \left(\frac{d}{w}\right) \text{H/m}$$

$$= \frac{1.26 \mu_r}{K_{L1}} \left(\frac{d}{w}\right) \mu\text{H/m} \quad (1)$$

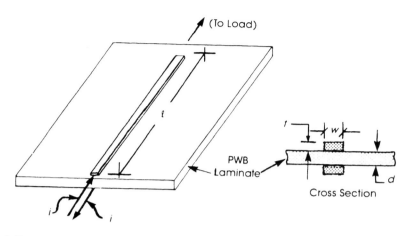

Figure 2.40 These two conductors represent lands on a printed wiring board. Current flows down the top land to the load and returns via the bottom land.

For air (or vacuum) where $\mu_r \approx 1$,

$$\frac{L}{l} \approx \frac{1.26}{K_{L1}} \left(\frac{d}{w}\right) \mu H/m \tag{2a}$$

$$= \frac{0.032}{K_{L1}} \left(\frac{d}{w}\right) \mu H/in \tag{2b}$$

Example 1:

Calculate the self-inductance of two lands $w = 0.025''$ wide, separated by $d = 0.060''$ and 6" long. Compare this value with measured data.

Step 1: Determine $d/w = 0.060''/0.025'' = 2.4$, as shown in Fig. 2.41.

Step 2: Enter curve, Fig. 2.41, at $d/w = 2.4$.

Step 3: Calculate the self-inductance from Eq. (2b):

$$\frac{L}{l} \approx \frac{0.032}{K_{L1}} \left(\frac{d}{w}\right) \mu H/in$$

$$L = \frac{0.032}{3.3} \left(\frac{0.060''}{0.025''}\right) \times 6'' \; \mu H$$

$$= 0.140 \; \mu H \; versus \; 0.145 \; \mu H \; \text{measured} \tag{3}$$

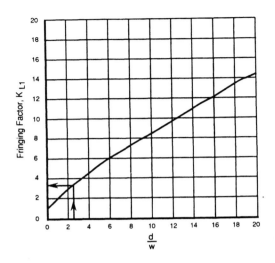

Figure 2.41 The fringing factor, K_{L1}, equals 3.3 for $d/w = 2.4$.

Derivation:

Formula Set C-6 sets the stage for Formula Sets L-6, C-4, and L-4 by:

1. Determining the fringing factors K_{L1} and K_{C1}.
2. Showing that the flux pattern for the upper half of two vertical, flat conductors is identical to that for a single flat conductor over a ground plane. Thus, if we make $d = 2h$, the correct fringing factor to use in this formula set is K_{L1}, that determined in Formula Set C-6.
3. If we replace d with $2h$ in Eq. (1), we get

$$\frac{L}{l} \approx \frac{\mu_r \mu_0}{K_{L1}} \left(\frac{2h}{w}\right) \text{H/m} \qquad (4)$$

which is twice the self-inductance per unit length from Eq. (1) in Formula Set L-6 because twice as many flux lines are generated for the same amount of current.

Commentary and Conclusions:

1. Equations (1), (2a), an (2b) showed that the inductance is independent of the relative dielectric constant, as we would expect. The inductance, however, depends inversely on the inductive fringing factor, K_{L1}, because the conductor geometry determines the magnetic flux distribution and K_{L1} is the factor that takes this into account.

2.3.5 Self-Inductance of Two Horizontal, Flat Conductors, L-5

The value of the self-inductance of a loop consisting of flat conductors (see Fig. 2.42) contributes to ground and power supply bus noise as was noted in Formula Set L-4.

Equations:

The self-inductance per unit length is given approximately by

$$\frac{L}{l} \approx \frac{\mu_r \mu_0}{\pi} \ln\left[\frac{\pi(d-w)}{w+t} + 1\right] \text{H/m} \tag{1}$$

$$= 0.4 \ln\left[\frac{\pi(d-w)}{w+t} + 1\right] \mu\text{H/m} \tag{2a}$$

$$= 0.01 \ln\left[\frac{\pi(d-w)}{w+t} + 1\right] \mu\text{H/in} \tag{2b}$$

For air (or vacuum) where $\mu_r \approx 1$ and $(w+t)/\pi d \ll 1$,

$$\frac{L}{l} \approx \frac{\mu_r \mu_0}{\pi} \ln\left(\frac{\pi d}{w+t}\right) \text{H/m} \tag{3}$$

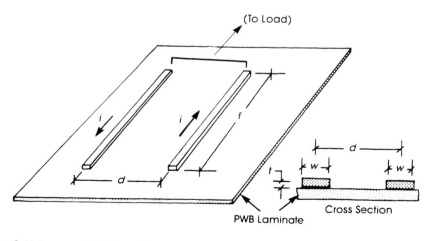

Figure 2.42 Two parallel, flat conductors represent lands on a printed wiring board. Current *i* flows down one conductor to the load and returns in the other.

$$= 0.4 \ln\left(\frac{\pi d}{w + t}\right) \mu\text{H/m} \qquad (4a)$$

$$= 0.01 \ln\left(\frac{\pi d}{w + t}\right) \mu\text{H/in} \qquad (4b)$$

Example 1:

Calculate the inductance for a loop formed by two parallel PWB lands that are each 0.025" wide, 6" long and spaced 1" apart. Insofar as 2 oz. foil is used, $t = 0.0028"$. Because the ratio $(w + t)/\pi d = (0.020" + 0.0028")/(\pi \times 1") = 0.007$ is $\ll 1$, Eq. (4b) is used. Thus,

$$L \approx 0.01 \ln\left(\frac{\pi d}{w + t}\right) \times l \; \mu\text{H} \qquad (5)$$

$$= 0.01 \ln\left(\frac{\pi \times 1"}{0.025" + 0.0028"}\right) \times 6" \mu\text{H} \qquad (6)$$

$$= 0.28 \; \mu\text{H}$$

Measurements made at 100 kHz yielded 0.33 μH *versus* 0.28 μH calculated above.

Example 2:

Assume now that the lands are now 0.065" apart. Recalculate the inductance using both Eqs. (4b) and (2b). Compare these values with results measured at 100 kHz:

$$\frac{L}{l} \approx 0.01 \ln\left(\frac{\pi d}{w + t}\right) \mu\text{H/in}$$

$$L \approx 0.01 \ln\left(\frac{\pi \times 0.065"}{0.025" + 0.0028"}\right) \times 6" \mu\text{H}$$

$$= 0.12 \; \mu\text{H} \; versus \; 0.14 \; \mu\text{H measured} \qquad (7)$$

From Eq. (2b),

$$\frac{L}{l} \approx 0.01 \ln\left[\frac{\pi(d - w)}{w + t} + 1\right] \mu\text{H/in} \qquad (8)$$

$$L \approx 0.01 \ln\left[\frac{\pi(0.065'' - 0.025'')}{0.025'' + 0.0028''} + 1\right] \times 6''\mu\text{H}$$

$$= 0.10 \ \mu\text{H} \ versus \ 0.14 \ \mu\text{H measured} \tag{9}$$

Derivation:

Paralleling the methods used in Formula Set C-5, the inductance per unit length is derived from Formula Set L-1 which gives the inductance per unit length for circular conductors as

$$\frac{L}{l} \approx \frac{\mu_r\mu_0}{\pi} \ln\left(\frac{d}{r}\right) \text{H/m} \tag{10}$$

As in Formula Set C-5, Eq. (10) is modified for rectangular conductors and close spacing considerations, and yields Eqs. (1) and (3):

$$\frac{L}{l} \approx \frac{\mu_r\mu_0}{\pi} \ln\left[\frac{\pi(d-w)}{w+t} + 1\right] \text{H/m}$$

$$\approx \frac{\mu_r\mu_0}{\pi} \ln\left(\frac{\pi d}{w+t}\right) \text{H/m}, \quad \text{for} \quad \frac{w+t}{\pi d} \ll 1$$

Commentary and Conclusions:

1. Favorable correlation with experimental results shows that the equations above represent valid approximations for the self-inductance of lands on printed wiring boards and other configurations using coplanar rectangular conductors. More exact solutions are possible, but these are beyond the scope and intent of this book.

2.3.6 Self-Inductance of a Long Flat Conductor and a Ground Plane, L-6

The value of self-inductance of a loop consisting of a long flat conductor and a ground plane shown in Fig. 2.43 is obtained by replacing d with h in Formula Set L-4 and properly applying the fringing factor, K_{L1}.

Equations:

The self inductance per unit length is defined by

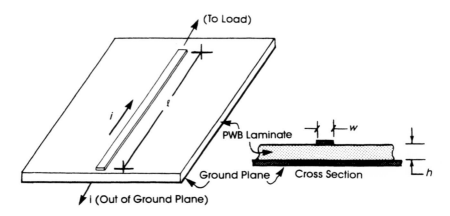

Figure 2.43 Current flows down the flat conductor to the load and returns via the ground plane, thus forming a loop.

$$\frac{L}{l} = \frac{\mu_r \mu_0}{K_{L1}} \left(\frac{h}{w}\right) \text{H/m} \tag{1}$$

For air (or vacuum), where $\mu_r \approx 1$,

$$\frac{L}{l} \approx \frac{1.26}{K_{L1}} \left(\frac{h}{w}\right) \mu\text{H/m} \tag{2a}$$

$$= \frac{0.032}{K_{L1}} \left(\frac{h}{w}\right) \mu\text{H/in} \tag{2b}$$

Example:

Calculate the self-inductance of the loop shown above for these dimensions: land width, $w = 0.015''$, height above the ground plane, $h = 0.060''$, and length, $l = 11.1''$. Compare calculated value with measured results.

Step 1: Determine $2h/w = (2 \times 0.060'')/0.015'' = 8$

Step 2: Determine the fringing factor, K_{L1}, from Fig. 2.44.

Step 3: Calculate the self-inductance from Eq. (2b)

$$L = \frac{0.032}{7.25} \times \left(\frac{0.060''}{0.015''}\right) \times 11.1'' \ \mu\text{H}$$

$$= 0.20 \ \mu\text{H} \ versus \ 0.22 \ \mu\text{H measured} \tag{3}$$

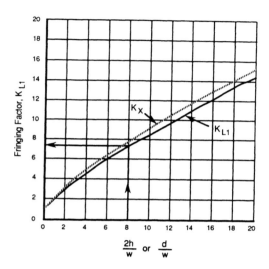

Figure 2.44 Plotted from values listed in Table 2.2, this curve shows the inductive fringing factor, K_{L1}, as a function of $2h/w$. An extra curve showing the "fringing factor" for round conductors, K_x has been added. Please see Comment 1. For the example, enter curve at $2h/w = 8$. Find $K_{L1} = 7.25$.

Commentary and Conclusions:

1. Figure 2.44 shows the fringing factor, K_x, for round conductors. This is obtained by letting the perimeter of each of the round conductors equal that of the rectangular conductors, and comparing the inductance obtained to that of a parallel plate inductor without fringing. Thus,

$$K_x = \frac{\pi}{\ln\left(\frac{\pi d}{w}\right)} \times \left(\frac{d}{w}\right) \tag{4}$$

2. As in previous formula sets, the self-inductance of a loop formed by a conductor that is h units over a ground plane is $1/2$ that for two conductors spaced $2h$ units apart because only $1/2$ the flux lines are generated.
3. As discussed in Formula Set L-3, the addition of the ground plane reduces the self-inductance by only a factor of 2.
4. Substantial reductions in inductance can be made by increasing the conductor

width and reducing the height. If $w = 0.100''$ and $h = 0.015''$, the inductance is calculated to be (for 11.1" of length):

$$L = \frac{0.032}{1.5} \times \left(\frac{0.015''}{0.100''}\right) \times 11.1 \ \mu\text{H} \qquad (5)$$

$$= 0.036 \ \mu\text{H} \qquad (6)$$

This would be a 5.6 times improvement over the 0.015" wide land, spaced 0.060" above a ground plane.

2.3.7 Mutual Inductance between Two Flat Conductors Near a Ground Plane, L-7

The value of mutual inductance between two flat conductors over a ground plane is needed to calculate the inductive crosstalk between circuits on printed wiring boards. (See Fig. 2.45.)

Equations:

The mutual inductance per unit length is given approximately by

$$\frac{L_m}{l} \approx \frac{\mu_r \mu_0}{4\pi} \ln\left[1 + \left(\frac{2h}{d}\right)^2\right] \text{H/m} \qquad (1)$$

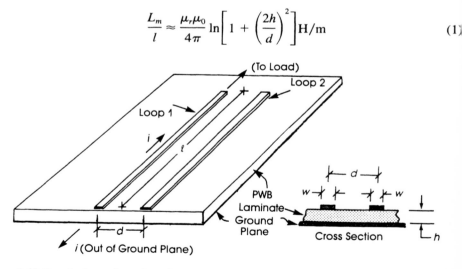

Figure 2.45 Two lands are located on the top layer of a printed wiring board with a ground plane on the bottom. Current, i, flows down land 1 to a load and returns via the ground plane. A voltage, due to a time rate of change of the current in the land 1-to-ground-plane loop, is induced in the circuit formed by land 2 and the ground plane.

$$\approx \frac{\mu_r \mu_0}{\pi} \left(\frac{h}{d}\right)^2 \text{ H/m} \tag{2}$$

$$= 0.4 \left(\frac{h}{d}\right)^2 \mu\text{H/m, for } \mu_r = 1 \tag{3a}$$

$$= 0.01 \left(\frac{h}{d}\right)^2 \mu\text{H/in} \tag{3b}$$

Please see Formula Set C-7 for approximation accuracy regarding Eq. (2).

Example:

Calculate the mutual inductance between 2 parallel 0.020" lands, 4" long, 1" apart, and positioned 0.060" above a ground plane. Using Eq. (3b),

$$\frac{L_m}{l} \approx 0.01 \left(\frac{h}{d}\right)^2 \mu\text{H/in}$$

$$L_m \approx 0.01 \left(\frac{0.060"}{1"}\right)^2 \times 4" \ \mu\text{H}$$

$$= 0.000144 \ \mu\text{H}$$

$$= 0.144 \text{ nH} \tag{4}$$

Commentary and Conclusions:

1. Equation (1) of this formula set is the same as the corresponding equation in Formula Set L-3 (which is for circular conductors). In this formula set, the assumption is made that the total magnetic flux distribution is approximately the same for flat conductors as it is for circular conductors. Thus, the flux lines linking loop 2 due to the current in loop 1 will be approximately the same. Please see comment 1 in Formula Set L-6.

2.3.8 Four-Conductor-System Mutual Inductance, L-8

This formula set has wide application for determining the mutual inductance between wiring cable pairs or land pairs on printed circuit boards (see Fig. 2.46). Coexistence between high and low signal circuits can occur with minimum crosstalk on the same board, provided that their respective center lines are orthogonal. See EXP L-8 for additional details.

Note: The two circuits are assumed to be ground and mutually isolated.

Introduction:

Formula Set L-7 and L-10 solve for the mutual inductance between two conductors in the presence of ground planes. This formula set solves for the mutual inductance between two parallel line sets. Effects of shielding are neglected and the rectangular conductor sets are assumed to have the same mutual inductance as round conductors. The examples illustrate an important (and perhaps surprising) result: the mutual inductance can be positive, negative or even *zero*. Thus, slight changes in cable routing can produce significantly different crosstalk levels.

Equations:

From Ref. [1], the mutual inductance is given by

$$\frac{L_m}{l} = \frac{\mu_r \mu_0}{2\pi} \ln\left(\frac{D_{14} \times D_{23}}{D_{13} \times D_{24}}\right) \text{H/m} \tag{1}$$

$$= 0.2 \ln\left(\frac{D_{14} \times D_{23}}{D_{13} \times D_{24}}\right) \mu\text{H/m}, \quad \text{for } \mu_r = 1 \tag{2a}$$

$$= 0.005 \ln\left(\frac{D_{14} \times D_{23}}{D_{13} \times D_{24}}\right) \mu\text{H/in} \tag{2b}$$

Example 1a:

Calculate the mutual inductance between high level circuit conductors, lands 1 and 2, spaced 0.5" apart, placed 0.060" directly above two low-level signal conductors 3 and 4, as shown in Fig. 2.47. All lands are 6" long.

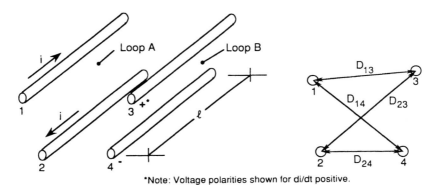

*Note: Voltage polarities shown for di/dt positive.

Figure 2.46 Two conductor pairs with a time-varying current flowing in loop A (conductors 1 and 2) induce a voltage in loop B (conductors 3 and 4) that is proportional to the mutual inductance between the two pairs.

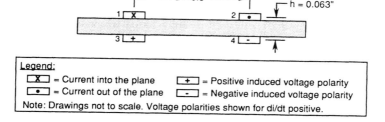

Figure 2.47 Land pair 1 and 2 form a loop in which a time-varying current flows down conductor 1 to the load and returns via conductor 2. A voltage is induced in the loop formed by land pair 3 and 4.

Using Eq. (2b),

$$\frac{L_m}{l} = 0.005 \ln\left(\frac{D_{14} \times D_{23}}{D_{13} \times D_{24}}\right) \mu\text{H/in}$$

$$D_{14} = D_{23} = \sqrt{d^2 + h^2} \tag{3}$$

$$D_{13} = D_{24} = h \tag{4}$$

Thus,

$$L_m = 0.005\left[\ln\left(\frac{d^2 + h^2}{h^2}\right)\right] \times l\ \mu\text{H}$$

$$= 0.005\left\{\ln\left[1 + \left(\frac{d}{h}\right)^2\right]\right\} \times l\ \mu\text{H}$$

$$= 0.005\left\{\ln\left[1 + \left(\frac{0.5''}{0.063''}\right)^2\right]\right\} \times 6''\mu\text{H}$$

$$= 0.125\ \mu\text{H}$$

$$= 125\ \text{nH} \tag{5}$$

Example 1b:

Figure 2.48 shows the same configuration as Example 1a except the signal land set is moved 4" to the right.

In this case,

$$D_{14} = \sqrt{(d_1 + d_2)^2 + h^2} \quad (6)$$

$$D_{23} = \sqrt{(d_2 - d_1)^2 + h^2} \quad (7)$$

$$D_{13} = D_{24} = \sqrt{d_2^2 + h^2} \quad (8)$$

For $l = 6''$,

$$L_m = 0.030 \ln\left[\frac{\sqrt{(d_1 + d_2)^2 + h^2}\sqrt{(d_2 - d_1)^2 + h^2}}{d_2^2 + h^2}\right]$$

$$= 0.030 \ln\left[\frac{\sqrt{(4.5'')^2 + (0.063'')^2}\sqrt{(3.5'')^2 + (0.063'')^2}}{(4'')^2 + (0.063'')^2}\right]$$

$$= -0.00047 \ \mu H$$

$$= -0.47 \ nH \quad (9)$$

If the conductors 3 and 4 are moved to the top of the board, $h = 0$ and L_m is

$$L_m = 0.030 \ln\left[1 - \left(\frac{0.5''}{4''}\right)^2\right]$$

$$= -0.47 \ nH, \text{ the same as (9) because } h \ll d_1, d_2 \quad (10)$$

Comments:

1. The minus sign indicates a reversal in voltage polarity in the "receiving" land circuit.
2. Moving the land sets 4″ apart reduces the crosstalk by a factor of 264 (48 dB). This example illustrates the benefit of separating high and low level circuits.

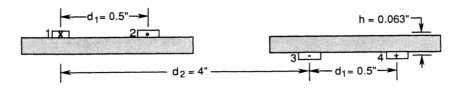

Figure 2.48 This structure yields a much lower mutual inductance than that shown in Fig. 2.47 with a corresponding decrease in crosstalk.

Example 2:

Figure 2.49 shows the same land configuration as Fig. 2.47, except that the lands carrying the high-level signals are placed one on top of the other as are the receiving lands.

The distances between conductors are

$$D_{14} = D_{23} = \sqrt{d^2 + h^2} \tag{11}$$

$$D_{13} = D_{24} = d \tag{12}$$

The mutual inductance thus is

$$L_m = 0.005\left[\ln\left(\frac{d^2 + h^2}{d^2}\right)\right] \times l \ \mu H$$

$$= 0.005\left\{\ln\left[1 + \left(\frac{h}{d}\right)^2\right]\right\} \times l \ \mu H$$

$$= 0.005\left\{\ln\left[1 + \left(\frac{0.063''}{0.5''}\right)^2\right]\right\} \times 6'' \ \mu H$$

$$= 0.00047 \ \mu H$$

$$= 0.47 \ nH \tag{13}$$

Comment:

1. Example 2 shows the very large benefits derived by close attention to printed circuit board layout. A $125/0.47 = 266 \times$ reduction in crosstalk is achieved by simply rearranging the high and low level circuits.

Example 3:

Transposition and division of the high level conductors yields even greater benefits. In fact, this arrangement has been patented. Please see Ref. [2]. Figure 2.50 illustrates the land set pattern.

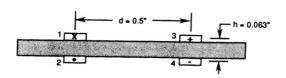

Figure 2.49 The "transmitting" loop is located on the left, the "receiving" loop is on the right.

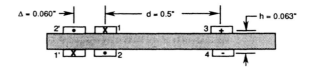

Figure 2.50 The high-level current is divided into two paths, land 1 and 1', with the return curren flowing in lands 2 and 2'.

The total induced voltage $e(t)$ in the receiving circuit 3–4 is given by

$$e(t) = L_m \frac{di_1(t)/2}{dt} + L'_m \frac{di_1(t)/2}{dt}$$

$$= \frac{L_m + L'_m}{2} \frac{di_1(t)}{dt}$$

$$= L_{mt} \frac{di_1(t)}{dt} \tag{14}$$

For land sets 1–2 and 3–4, the mutual inductance is given by Eq. (15):

$$L_m = 0.005 \left[\ln\left(\frac{d^2 + h^2}{d^2}\right) \right] \times l \, \mu\text{H}$$

$$= 0.005 \left\{ \ln\left[1 + \left(\frac{h}{d}\right)^2 \right] \right\} \times l \, \mu\text{H} \tag{15}$$

For land sets 1'–2' and 3–4, d in Eq. (14) is replaced by $d + \Delta$ (where Δ is a distance small compared with d) and a minus sign added because the current flow in the 1'–2' loop is negative with respect to that in the 1–2 loop.

$$L'_m = -0.005 \left\{ \ln\left[\frac{(d + \Delta)^2 + h^2}{(d + \Delta)^2} \right] \right\} \times l \, \mu\text{H}$$

$$= -0.005 \left\{ \ln\left[1 + \left(\frac{h}{d + \Delta}\right)^2 \right] \right\} \times l \, \mu\text{H} \tag{16}$$

Using Eqs. (14), (15), and (16) to get the total mutual inductance:

$$L_{mt} = 0.0025 \left\{ \ln\left[1 + \left(\frac{h}{d}\right)^2 \right] - \ln\left[1 + \left(\frac{h}{d + \Delta}\right)^2 \right] \right\} \times l \, \mu\text{H} \tag{17}$$

For $d = 0.5''$, $h = 0.063''$, $\Delta = 0.06''$, and $l = 6''$, we get

$$L_{mt} = 0.0025\left\{\ln\left[1 + \left(\frac{0.063''}{0.5''}\right)^2\right] - \ln\left[1 + \left(\frac{0.063''}{0.56''}\right)^2\right]\right\} \times 6'' \, \mu\text{H}$$

$$= 0.047 \text{ nH} \qquad (18)$$

If $h \ll d$, a common case, the logarithmic terms can be series expanded

$$\ln(1 + x) = x - \frac{x^2}{2} + \frac{x^3}{3} - \ldots \text{ for } -1 < x \leq +1 \qquad (19)$$

$$L_{mt} \approx 0.0025\left[\left(\frac{h}{d}\right)^2 - \left(\frac{h}{d + \Delta}\right)^2\right] \times l \, \mu\text{H} \qquad (20)$$

For $\Delta \ll d$, Eq. (20) reduces to

$$L_{mt} \approx 0.005\left(\frac{h}{d}\right)^2\left(\frac{\Delta}{d}\right) \times l \, \mu\text{H} \qquad (21)$$

We can compare the results of this approximation to that using Eq. (18).

$$L_{mt} \approx 0.005\left(\frac{0.063''}{0.5''}\right)^2\left(\frac{0.060''}{0.5''}\right) \times 6'' \, \mu\text{H}$$

$$= 0.057 \text{ nH (about 1.7 dB higher)} \qquad (22)$$

Example 4:

Figure 2.51 shows yet another configuration which, surprisingly, yields the same results as Example 3.

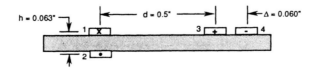

Figure 2.51 The two receiving lands are located on the right and are closely spaced Δ'' apart.

Repeating Eq. (2b),

$$\frac{L_m}{l} = 0.005 \ln\left(\frac{D_{14} \times D_{23}}{D_{13} \times D_{24}}\right) \mu\text{H/in}$$

The distances are

$$D_{14} = d + \Delta \qquad D_{23} = \sqrt{d^2 + h^2}$$
$$D_{13} = d \qquad D_{24} = \sqrt{(d + \Delta)^2 + h^2}$$

Applying these to Eq. (2b),

$$\frac{L_m}{l} = 0.005 \ln\left[\frac{(d + \Delta)\sqrt{d^2 + h^2}}{d\sqrt{(d + \Delta)^2 + h^2}}\right] \mu\text{H/in} \tag{23}$$

We can square the quantity inside the logarithm by dividing the right-hand side of Eq. (23) by 2, and simplifying:

$$\frac{L_m}{l} = 0.0025 \ln\left[\frac{(d + \Delta)^2(d^2 + h^2)}{d^2((d + \Delta)^2 + h^2)}\right] \mu\text{H/in} \tag{24}$$

$$= 0.0025 \ln \frac{\left[1 + \left(\frac{h}{d}\right)^2\right]}{\left[1 + \left(\frac{h}{d + \Delta}\right)^2\right]} \mu\text{H/in} \tag{25}$$

$$= 0.0025 \left\{\ln\left[1 + \left(\frac{h}{d}\right)^2\right] - \ln\left[1 + \left(\frac{h}{d + \Delta}\right)^2\right]\right\} \mu\text{H/in} \tag{26}$$

which is equivalent to Eq. (17).

Comments:

1. The invention status of this configuration is unknown.
2. As seen above, we get the same crosstalk reduction benefits as Example 4 for these particular semi-coplanar geometries. If the lands are not coplanar, it is recommended that the mutual inductance be calculated for each configuration and compared.

Example 5:

This example illustrates the principle of centerline orthogonality, as shown in Fig. 2.52.

Again repeating Eq. (2b),

$$\frac{L_m}{l} = 0.005 \ln\left(\frac{D_{14} \times D_{23}}{D_{13} \times D_{24}}\right) \mu\text{H/in}$$

From the figure, that

$$D_{13} = D_{14} \text{ and } D_{24} = D_{23}$$

is apparent. Thus,

$$\frac{L_m}{l} = 0.005 \ln\left(\frac{D_{14} \times D_{23}}{D_{14} \times D_{23}}\right) \mu\text{H/in}$$

$$= 0.005 \ln 1$$

$$= 0 \ \mu\text{H/in} \tag{27}$$

Comment:

1. Please note that we will get the same result by placing conductor pair 1 and 2 anywhere on the orthogonal centerline even directly below conductor pair 3 and 4.
2. Any combination of distances that makes the ratio $(D_{14} \times D_{23})/(D_{13} \times D_{24}) = 1$ theoretically produces zero mutual inductance.

Example 6:

Placing the receiving conductors between the transmitting conductors as in Fig. 2.53 intuitively suggests an increase in mutual inductance.

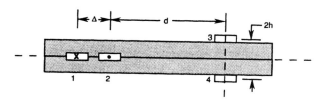

Figure 2.52 This is a multilayer board where the transmitting conductors are located on layer 2 and the receiving on layers 1 and 3, respectively.

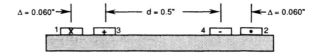

Figure 2.53 Here the receiving loop is located inside the transmitting loop producing, as expected, a large mutual inductance.

The distances are

$$D_{14} = D_{23} = 0.56'' \text{ and } D_{13} = D_{24} = 0.06''$$

Calculating the mutual inductance for a 6" land length,

$$L_m = 0.005 \ln\left(\frac{0.56'' \times 0.56''}{0.060'' \times 0.060''}\right) \times 6'' \; \mu\text{H}$$

$$= 134 \text{ nH} \qquad (28)$$

which is about the same as that found in Example 1b and substantially larger than the geometries designed for minimizing mutual inductance.

Commentary and Conclusions:

1. The number of possible geometric combinations, of course, is infinite. The examples shown in this formula set illustrate only a few of them. However, from these, we can see how the land arrangement influences the mutual inductance values and avoids those which produce undesirable results.

REFERENCES

1. Rogers, W.E., *Introduction to Electric Fields*, New York, McGraw-Hill, 1950, p. 316.
2. Vogelsberg, Patent #24 45 534, Federal Republic of Germany.

2.3.9 Self-Inductance of a Stripline (Flat Conductor between Two Ground Planes), L-9

Following the methods used in Formula Set C-9, the value of the self-inductance per unit length of a long flat conductor and two ground planes (stripline) is derived from L-6 (see Fig. 2.54). But in this case, the value calculated from L-6 is *divided* by two and a new fringing factor, K_{L2}, is used. As in Formula Set C-9, excellent correlation with data from Ref. [1] is shown.

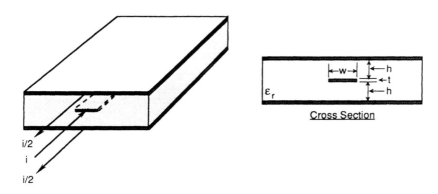

Figure 2.54 A flat conductor lies between two ground planes. A current, i, flows down the conductor and returns via the ground planes, $i/2$ in each.

Equations:

The self-inductance per unit length is defined as

$$\frac{L}{l} \approx \frac{\mu_r \mu_0}{2K_{L2}} \left(\frac{h}{w}\right) \text{H/m} \qquad (1)$$

$$= \frac{0.63}{K_{L2}} \left(\frac{h}{w}\right) \mu\text{H/m} \qquad (2a)$$

$$= \frac{0.016}{K_{L2}} \left(\frac{h}{w}\right) \mu\text{H/in} \qquad (2b)$$

Example:

Using Eq. (2a), determine the inductance per unit length for the configuration shown in Fig. 2.55. Compare this value with that shown in Ref. [1].

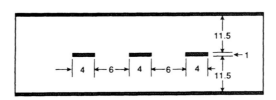

Figure 2.55 A computer-generated finite-element analysis was used to determine the self-inductance of loops formed by each land and the ground planes for this structure.

Equation (2a) gives the self-inductance per unit length as

$$\frac{L}{l} \approx \frac{0.63}{K_{L2}} \left(\frac{h}{w}\right) \text{H/m}$$

From Fig. 2.55, $h = 11.5$, $w = 4$, and therefore, $2h/w = 5.75$. K_{L2} from Fig. 2.56 is then 4.0.

Calculating the self-inductance per unit length from Eq. (2a):

$$\frac{L}{l} = \frac{0.63}{4.0} \times \frac{11.5}{4}$$

$$= 0.453 \ \mu\text{H/m}$$

$$= 0.00453 \ \mu\text{H/cm}$$

$$= 4.53 \text{ nH/cm } versus \text{ } 4.71 \text{ nH/cm shown in Ref. [1].} \tag{3}$$

Commentary and Conclusions:

1. Following the methods of Formula Set C-9, the value for the self-inductance for a flat conductor between two ground planes is obtained from Formula Set L-6. Here we have two inductors in parallel, one formed by the conductor and

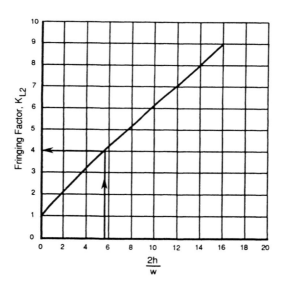

Figure 2.56 For $2h/w = 5.75$, $K_{L2} = 4.0$.

the bottom ground plane, the other by the conductor and the top ground plane. Thus, the inductance value for this formula set, L-9, will be approximately one-half that found in Formula Set L-6, as modified by the new fringing factor K_{L2}.

2. As in Formula Set C-9, we see that this relatively easy method of determining the inductance has an excellent correlation with Ref. [1].

REFERENCES

1. Olsen, L.T., "Application of the Finite Element Method to Determine the Electrical Resistance, Inductance, Capacitance Parameters for the Circuit Package Environment," *Trans. IEEE*, Vol. CHMT-5, No. 4, December 1982, pp. 486–492.

2.3.10 Mutual Inductance between Two Flat Conductors Near Two Ground Planes, L-10

The value of the mutual inductance per unit length between two long, flat conductors near two ground planes, as shown in Fig. 2.57, is derived from Formula Sets C-10 and Z_0-(ALL). The formulas presented here provide values which have excellent correlation with finite element analysis results.

Equations:

The mutual inductance per unit length is given approximately by

$$\frac{L_m}{l} \approx \frac{\mu_r\mu_0}{4\pi}\left(\frac{h}{d}\right)^2 \text{ H/m} \tag{1}$$

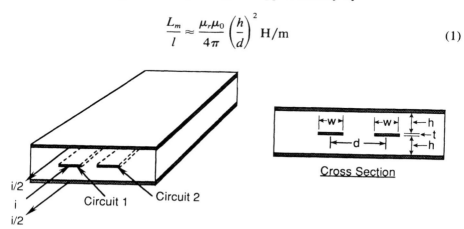

Figure 2.57 A time-varying current in circuit 1 flows down the conductor to the load and returns via the two ground planes, 1/2 in each. This current produces a voltage in circuit 2 that is directly proportional to the mutual inductance between the two circuits.

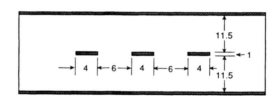

Figure 2.58 The three lands, located between two ground planes, each form self-inductances with the ground planes and mutual inductances with each other. The ground planes are assumed to be connected.

$$= 0.1\left(\frac{h}{d}\right)^2 \ \mu H/m \tag{2a}$$

$$= 0.00254\left(\frac{h}{d}\right)^2 \ \mu H/in \tag{2b}$$

Example:

Calculate the value of mutual inductance between lands 1 and 2 as shown in Fig. 2.58. Compare this value with that from Ref. [1].

The mutual inductance per unit length between lands 1 and 2 is from Eq. (2a):

$$\frac{L_m}{l} \approx 0.1\left(\frac{h}{d}\right)^2 \ \mu H/m$$

$$= 0.1\left(\frac{11.5}{10}\right)^2 \ \mu H/m$$

$$= 1.32 \ nH/cm \ versus \ 1.15 \ nH/cm \ \text{from Ref. [1]}. \tag{3}$$

Derivation:

Formula Set C-3 gives the relationship between mutual inductance, mutual capacitance, and conductor-to-ground plane characteristic impedance *for homogeneous media* as

$$\frac{L_m}{l} = Z_{01}Z_{02}\frac{C_m}{l} \tag{4}$$

From Formula Set Z_0-(ALL), the characteristic impedance for a loop consisting of a flat conductor between two ground planes as

$$Z_0\text{-}9 = \frac{1}{2K_{C2}}\left(\frac{h}{w}\right)\sqrt{\frac{\mu_r\mu_0}{\varepsilon_r\varepsilon_0}} \tag{5}$$

Because lands 1 and 2 are the same distance from the ground planes:

$$Z_{01} = Z_{02} \qquad (6)$$

Thus,

$$(Z_{01})^2 = \frac{\mu_r \mu_0}{4\varepsilon_r \varepsilon_0 K_{C2}^2} \left(\frac{h}{w}\right)^2 \qquad (7)$$

From C-10:

$$\frac{C_m}{l} \approx \frac{\varepsilon_r \varepsilon_0}{\pi} K_{C2}^2 \left(\frac{w}{d}\right)^2 \text{ F/m} \qquad (8)$$

Combining Eqs. (4), (6), (7) and (8), we get Eq. (1), the expression for the mutual inductance:

$$\frac{L_m}{l} \approx \frac{\mu_r \mu_0}{4\pi} \left(\frac{h}{d}\right)^2 \text{ H/m}$$

Commentary and Conclusions:

1. The example shows good correlation with the value shown in Ref. [1].

REFERENCES

1. Olsen, L.T., "Application of the Finite Element Method to Determine the Electrical Resistance, Inductance, Capacitance Parameters for the Circuit Package Environment," *Trans. IEEE*, Vol. CHMT-5, No. 4, December 1982, pp. 486–492.

2.3.11 Self-Inductance of Coaxial Cables, L-11

Capacitance per foot and characteristic impedance are usually listed in coaxial cable manufacturers' catalogs. Self-inductance per foot is sometimes listed, but not on all occasions.

This Formula Set presents equations for coaxial cable self-inductance. Calculated values are compared with those derived from Z_0 for several commercially available cables. Figure 2.59 illustrates a coaxial cable and shows the dimensions of interest.

Equations:

The self-inductance per unit length of a long coaxial cable, neglecting inductance internal to the center conductor, is given by

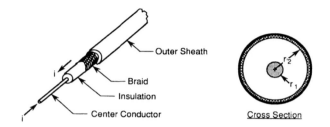

Figure 2.59 This is the same figure as used in Formula Set C-11. The self-inductance results from the loop formed by current, i, flowing down the center conductor and returning via the outer conductor.

$$\frac{L}{l} = \frac{\mu_r \mu_0}{2\pi} \ln\left(\frac{r_2}{r_1}\right) \text{H/m} \quad (1)$$

$$= 0.2\mu_r \ln\left(\frac{r_2}{r_1}\right) \mu\text{H/m} \quad (2a)$$

$$= 0.005\mu_r \ln\left(\frac{r_2}{r_1}\right) \mu\text{H/in} \quad (2b)$$

Example:

Calculate the self-inductance per foot for several commercially available cables. Compare these values with those derived from Z_0 and ε_r as listed in Ref. [1].

Equation (2b) is used to calculate the values. Type RG-6/U has $r_1 = 0.0185''$, $r_2 = 0.090''$, $\varepsilon_r = 1.64$, and $Z_0 = 75\ \Omega$.

$$L = 0.005 \ln\left(\frac{0.090''}{0.0185''}\right) \times 12'' \text{ pF}$$

$$= 0.095\ \mu\text{H/ft (for type RG-6/U)} \quad (3)$$

From the $LCRZ_0$ Analogy, we can show that

$$\frac{L}{l} = \frac{Z_0}{300}\sqrt{\mu_r \varepsilon_r}\ \mu\text{H/m} \quad (4)$$

For $\mu_r = 1$ and converting to $\mu\text{H/ft}$,

$$\frac{L}{l} = \frac{Z_0}{300}\sqrt{\varepsilon_r} \times \frac{12}{39.37}\ \mu\text{H/ft} \quad (5)$$

$$\frac{L}{l} = 0.001 \, Z_0 \sqrt{\varepsilon_r} \; \mu\text{H/ft}$$

$$= 0.001 \times 75 \times \sqrt{1.64}$$

$$= 0.096 \, \mu\text{H/ft (for type RG-6/U)} \tag{6}$$

Table 2.7 lists the results of these calculations.

Derivation:

In most coaxial cable designs, the dielectric between the conductors is homogeneous. If this is the case, we can use Eq. (7) from the $LRCZ_0$ Analogy, which gives the self-inductance per unit length in terms of the capacitance per unit length as

$$\frac{L}{l} = \frac{\mu_r \mu_0 \varepsilon_r \varepsilon_0}{\left(\dfrac{C}{l} \, \text{F/m}\right)} \; \text{H/m} \tag{7}$$

From Formula Set C-11,

$$\frac{C}{l} = \frac{2\pi\varepsilon_r\varepsilon_0}{\ln\left(\dfrac{r_2}{r_1}\right)} \; \text{F/m} \tag{8}$$

Combining Eq. (7) with (8), we get Eq. (1).

$$\frac{L}{l} = \frac{\mu_r\mu_0\varepsilon_r\varepsilon_0}{(2\pi\varepsilon_r\varepsilon_0)/\ln\left(\dfrac{r_2}{r_1}\right)}$$

$$= \frac{\mu_r\mu_0}{2\pi} \ln\left(\frac{r_2}{r_1}\right) \tag{9}$$

Table 2.7
Calculated Inductance per foot *versus* Values Derived from Z_0 for Coaxial Cables

Type	Z_0 (Ω)	ε_r^*	$2r_2$	$2r_1$	Calc. (μH/ft)	from Z_0 (μH/ft)
RG-6/U	75	1.64	0.180"	0.037"	0.095	0.096
RG-6A/U	75	2.3	0.185"	0.028"	0.113	0.116
9393	93	1.64	0.064"	0.010"	0.111	0.121
9252	50	2.3	0.096"	0.028"	0.074	0.077

*ε_r = 1.64 for cellular polyethylene, 2.3 for solid polyethylene

REFERENCES

1. "Master Catalog," *Belden Wire and Cable*, Richmond, IN, 1989, pp. 118–143 and p. 354.

2.3.12 Self-Inductance of Circular and Square Loops, L-12

In some instances, the separation of parallel conductors is not small compared with the length, an example being a square or a rectangle with approximately equal side dimensions.

This formula set gives the approximate values for the self-inductance of circular and square loops, as seen in Figs. 2.60 and 2.61. Experimental measurements show that the equations below yield reasonably accurate results.

Equations:

Reference [1] gives the self-inductance of a circular loop for $r \ll R$ approximately for $\mu_r = 1$ as:

$$L \approx \mu_0 R\left[\left(\ln \frac{8R}{r}\right) - 2\right] \text{H}, r \ll R \quad (1)$$

$$= 1.26R\left[\left(\ln \frac{8R}{r}\right) - 2\right] \mu\text{H (dimensions in meters)} \quad (2a)$$

$$= 0.032R\left[\left(\ln \frac{8R}{r}\right) - 2\right] \mu\text{H (dimensions in inches)} \quad (2b)$$

Example 1:

Calculate the inductance for a 12" diameter circular loop of #12 gauge wire ($r = 0.040"$) and compare value with measured data. Using Eq. (2b),

$$L = 0.032 \times 6" \times \left[\left(\ln \frac{8" \times 6"}{0.040"}\right) - 2\right] \mu\text{H}$$

$$= 0.98 \ \mu\text{H} \ versus \ 1.00 \ \mu\text{H measured} \quad (3)$$

Adding the results of Eq. (12), we get 1.03 μH.

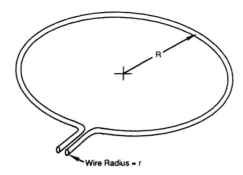

Figure 2.60 Circular loop of radius R with wire radius r.

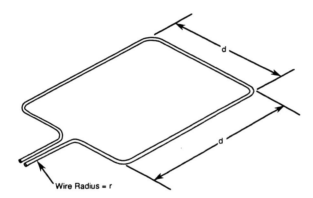

Figure 2.61 Square loop with sides d and wire with radius r.

Example 2:

Determine the dimensions of a square loop with the same area as the circular loop in Example 1. Compare the measured self-inductance value of this square loop with that calculated in Eq. (3). Calculate the square loop self-inductance using Eq. (2b), Formula Set L-1, and compare this value with the measured data.

1. The loop in Example 1 has an area = $\pi R^2 = 3.14 \times 6'' \times 6'' = 113$ in². A square with an equivalent area is 10.6" on the side. The measured value for this square is 1.08 μH, similar to the value found for the circular loop.
2. Formula Set L-1 gives the inductance for two long, parallel circular wires as

$$\frac{L}{l} \approx 0.01 \ln\left(\frac{d}{r}\right) \mu\text{H/in} \tag{4}$$

Putting in the dimensions of the 10.6" square,

$$L \approx 0.01 \ln\left(\frac{10.6"}{0.040"}\right) \times 10.6" \, \mu H$$

$$= 0.59 \, \mu H \tag{5}$$

Because Eq. (4) is for long conductors, the inductance of the end connections, short compared with the length, is ignored. In this case, the end connections are 10.6" long. If we multiply 0.59 μH by 2 to account for the flux generated by the end connections, we get 1.18 μH *versus* 1.08 μH measured.

Reference [2] gives the self-inductance of a square with $a \ll D$ for $\mu_r = 1$ as

$$L \approx 2\frac{\mu_0 D}{\pi}\left[\sinh^{-1}\left(\frac{D}{a}\right) - 1\right] H \text{ (dimensions in meters)} \tag{6}$$

Substituting $d = D$, $r = a$ and noting that

$$\sinh^{-1} x = \ln(x + \sqrt{x^2 + 1}) \tag{7}$$

we get for $r \ll d$,

$$L \approx 0.8d\left[\ln\left(\frac{2d}{r}\right) - 1\right] \mu H \text{ (dimensions in meters)} \tag{8a}$$

$$= 0.02 \, d\left[\ln\left(\frac{2d}{r}\right) - 1\right] \mu H \text{ (dimensions in inches)} \tag{8b}$$

With $d = 10.6"$ an d$r = 0.040"$, the inductance is

$$L \approx 0.02 \times 10.6" \times \left[\ln\left(\frac{2 \times 10.6"}{0.040"}\right) - 1\right] \mu H$$

$$= 1.12 \, \mu H \text{ *versus* } 1.08 \, \mu H \text{ measured} \tag{9}$$

Commentary and Conclusions:

1. Reference [1] derives an exact expression for the inductance of a circular loop for these assumptions: the loop is in free space ($\mu = \mu_0$), the current is concentrated on the wire axis and $r \ll R$ (the usual case). This exact solution requires the use of elliptic integrals. Equation (1) is only for the inductance

external to the loop. The author gives the complete approximate expression, including the internal inductance, L_i, for $\mu_r = 1$, as

$$L \approx L_e + L_i \qquad (10)$$

$$= \mu_0 R \left[\left(\ln \frac{8R}{r} \right) - 2 \right] + \frac{\mu_0 R}{4} \, H \qquad (11)$$

If we apply the second term in Eq. (11) to Example 1, we get

$$L_i = 0.032 \times \frac{6}{4}$$

$$= 0.048 \, \mu H \qquad (12)$$

which contributes only 5% to the value found in Eq. (3).

REFERENCES

1. Johnk, C.T.A., *Engineering Electromagnetic Fields and Waves*, New York, John Wiley and Sons, 1973, pp. 331–334.
2. Zahn, Markus, *Electromagnetic Field Theory*, Malabar, Florida, Robert E. Krieger, 1979, p. 343.

2.4 CHARACTERISTIC IMPEDANCE FOR VARIOUS GEOMETRIES, Z_0-(ALL)

Characteristic impedance, Z_0, is an intrinsic transmission line property used in defining high frequency termination requirements. For example, in order to prevent line reflections, a 75 Ω television receiver coaxial input line should be driven with a 75 Ω source impedance, and terminated with a 75 Ω load. Due to its importance in communications, Z_0 is often listed in handbooks where values of the capacitance or inductance per unit length is not.

The $LRCZ_0$ Analogy, Section 1.4, showed that, if we know Z_0, we can determine the capacitance or the inductance per unit length provided that the medium surrounding (which includes between) the conductors is homogeneous. Section 1.4 also showed that if the medium is not homogeneous, we needed to use a factor K to account for this condition. This formula set contains characteristic impedance values, with $\mu_r = 1$, for conductor configurations appearing in this book.

Basic Equations:

By definition, the characteristic impedance is related to inductance, L, and capacitance, C, (or alternatively, the inductance per unit length, L/l, and the capacitance per unit length, C/l) for lossless configurations by

$$Z_0 = \sqrt{\frac{L}{C}} \; \Omega \tag{1}$$

$$= \sqrt{\frac{L/l}{C/l}} \; \Omega \tag{2}$$

Z_0 Equations:

Z_0-1 long parallel circular conductors:

$$Z_0\text{-}1 = \frac{120}{\sqrt{\varepsilon_{r(\text{eff})}}} \ln\left\{\frac{d}{2r}\left[1 + \sqrt{1 - \left(\frac{2r}{d}\right)^2}\right]\right\} \Omega \tag{3}$$

$$\approx \frac{120}{\sqrt{\varepsilon_{r(\text{eff})}}} \ln\left(\frac{d}{r}\right) \Omega, \text{ for } \frac{2r}{d} \ll 1 \tag{4}$$

Comment:

Insulated wire pairs are common examples of long, parallel circular conductors. The insulation has a dielectric constant, ε_r, but since the insulation does not occupy the complete volume surrounding the conductors, the effective dielectric constant, $\varepsilon_{r(\text{eff})}$ will be less than ε_r. The inductance of the wire pairs, however, does not depend on the insulation. Please see Formula Sets C-1 and L-1 for further details.

Z_0-2 long circular conductor over a ground plane:

$$Z_0\text{-}2 = \frac{60}{\sqrt{\varepsilon_{r(\text{eff})}}} \ln\left\{\frac{h}{r}\left[1 + \sqrt{1 - \left(\frac{r}{h}\right)^2}\right]\right\} \Omega \tag{5}$$

$$\approx \frac{60}{\sqrt{\varepsilon_{r(\text{eff})}}} \ln\left(\frac{2h}{r}\right) \Omega, \text{ for } \frac{r}{h} \ll 1 \tag{6}$$

Comment:

(Same as for Z_0-1).

Formula Sets C-3 and L-3 are for mutual capacitances and inductances, hence characteristic impedance does not apply for these configurations.

Z_0-4 long parallel vertical flat conductors:

$$Z_0\text{-}4 = \frac{120\pi}{\sqrt{K_{L1}K_{C1}}\sqrt{\varepsilon_r}}\left(\frac{d}{w}\right)\Omega \tag{7}$$

Figures 2.16 and 2.30 plot K_{C1} and K_{L1}, respectively.

Z_0-5 long parallel horizontal flat conductors:

$$Z_0\text{-}5 \approx \frac{120}{\sqrt{\varepsilon_{r(\text{eff})}}}\ln\left(\frac{\pi(d-w)}{w+t}+1\right)\Omega \tag{8}$$

$$\approx \frac{120}{\sqrt{\varepsilon_{r(\text{eff})}}}\ln\left(\frac{\pi d}{w+t}\right)\Omega \tag{9}$$

Z_0-6 long flat conductor over a ground plane:

$$Z_0\text{-}6 \approx \frac{120\pi}{\sqrt{K_{L1}K_{C1}}\sqrt{\varepsilon_r}}\left(\frac{h}{w}\right)\Omega \tag{10}$$

Figures 2.16 and 2.30 plot K_{C1} and K_{L1}, respectively. Figure 2.62, which follows the equations, plots Z_0-6 versus $2h/w$ for $\varepsilon_r = 1$ and 4.5.

Formula Sets C-7, 8, 10 and L-7, 8, 10 are for mutual capacitances and inductances. C-12 and L-12 are for spheres and inductance loops, respectively.

Z_0-9 long flat conductor between two ground planes:

$$Z_0\text{-}9 \approx \frac{60\pi}{K_{C2}\sqrt{\varepsilon_r}}\left(\frac{h}{w}\right)\Omega \tag{11}$$

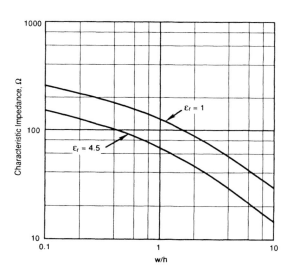

Figure 2.62 Microstrip characteristic impedance, Z_0-6, *versus* w/h for $\varepsilon_r = 1$ and $\varepsilon_r = 4.5$. Values are from Table 2.2, Formula Set C-6.

Figure 2.30 plots $K_{L2} = K_{C2}$ *versus* $2h/w$. Figure 2.29 plots the characteristic impedance *versus* conductor width, w for a particular stripline geometry.

Z_0-11 coaxial cables:

$$Z_0\text{-}11 = \frac{60}{\sqrt{\varepsilon_r}} \ln\left(\frac{r_2}{r_1}\right) \Omega \qquad (12)$$

Formula Set L-11 lists the characteristic impedance for several commercially available cables.

Commentary and Conclusions:

1. As noted, the above equations assume that $\mu_r = 1$. In the rare case where $\mu_r > 1$, considerations beyond the scope of this book must be made to determine the effect on characteristic impedance and inductance.
2. Care should be taken to consider properly the effect of dielectrics around or near the conductors. Each of the capacitive formula sets discusses the effective dielectric constant, ε_r, as appropriate.

Chapter 3
Crosstalk Analysis

3.1 INTRODUCTION

Crosstalk and noise represent one of the hardest problems to solve in electrical engineering. Almost without exception, electrical design projects have experienced the deleterious effects of unwanted interference signals.

This chapter provides the background information necessary to understand common causes of crosstalk and provides solutions for typical situations. Values are quantified by the application of formula sets.

Chapter 6, Experiments and Test Data, will describe 11 experiments specifically designed to illustrate crosstalk problems.

3.2 CAPACITIVE CROSSTALK

The following two subsections present crosstalk analyses for circuit board geometries. Subsection 3.2.1 investigates crosstalk between three circuit channels. Subsection 3.2.2 shows that crosstalk is significantly reduced by the use of a ground plane. (Chapter 6 describes test results for each of these two cases.)

3.2.1 Capacitive Coupling to Summing Junctions, CTC-1

The operational amplifier's summing-junction point and associated leads are very susceptible to crosstalk and noise pickup because capacitively coupled current is converted directly into amplifier output voltage. This analysis quantifies the crosstalk level from the printed wiring board layout configuration.

Analysis:

Consider the circuit diagram in Fig. 3.1 representing the input stage of an amplifier with gain G.

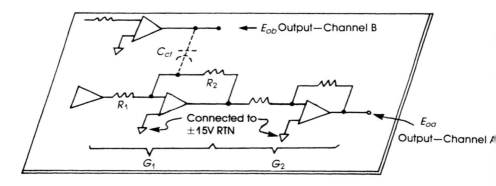

Figure 3.1 Two circuit channels are located near one another on the same circuit card or on a separate card in the electronics assembly. C_{ct} is the "stray" capacitance between the output of another amplifier (channel B) and the summing junction of the input stage of the subject channel A.

The crosstalk, in dB, between channels A and B is defined as the ratio of the output of channel A, with no signal input, divided by the output of channel B:

$$\text{crosstalk} = K = 20 \log \left| \frac{E_{oa}}{E_{ob}} \right| \tag{1}$$

Expressed in words, crosstalk is the voltage at the output of channel A due to a signal output from channel B.

The crosstalk between channel B and channel A can be computed as follows: The output of channel A is

$$E_{oa} = -j \frac{R_2}{X_{ct}} G_2 E_{ob} = -j 2\pi f R_2 C_{ct} G_2 E_{ob} \tag{2}$$

where

X_{ct} = reactance of crosstalk capacitance, C_{ct}

Because

$$G = G_1 G_2 \text{ and } G_1 = \frac{R_2}{R_1} \tag{3,4}$$

Then,

$$G_2 = G\frac{R_1}{R_2} \tag{5}$$

Substituting G_2 into Eq. (2) and rearranging yields the capacitive crosstalk:

$$\frac{E_{oa}}{E_{ob}} = -j2\pi f R_1 C_{ct} G \tag{6}$$

On a magnitude basis and converting to dB:

$$K_c = \text{crosstalk (dB)} = 20 \log \left|\frac{E_{oa}}{E_{ob}}\right|$$
$$= 20 \log (2\pi f R_1 C_{ct} G) \text{ dB} \tag{7}$$

Equation (7) is rearranged to solve for the maximum permissible value for C_{ct} to yield the specified crosstalk level:

$$20 \log (2\pi f R_1 C_{ct} G) \leq K_c \tag{8}$$

Taking the antilogarithm of Eq. (8), we get

$$2\pi f R_1 C_{ct} G \leq \log^{-1}\left(\frac{K_c}{20}\right) \tag{9}$$

or

$$C_{ct} \leq \frac{10^{12}}{2\pi f R_1 G} \log^{-1}\left(\frac{K_c}{20}\right) \text{pF} \tag{10}$$

Alternatively, we can compute C_{ct} if the ratio $|E_{oa}/E_{ob}|$ is known:

$$C_{ct} \leq \frac{10^{12}}{2\pi f R_1 G} \times \left|\frac{E_{oa}}{E_{ob}}\right| \text{pF} \tag{11}$$

Example:

A printed wiring board has 10 layers each 0.007" thick. As is common practice, all of the components are mounted on the top side of the board. In this example, we are focusing on the voltage induced in the channel A (victim) circuit by channels B and C, as shown in Fig. 3.2.

The Schematic:

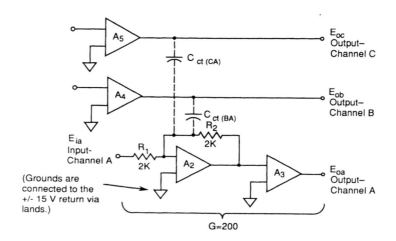

Figure 3.2 Three channels are shown here. Channel A is the "victim" circuit. Crosstalk is coupled between channels B-A and C-A through capacitances $C_{ct(BA)}$ and $C_{ct(CA)}$, respectively.

Crosstalk Specifications:

1. The ratio of the output of channel A due to channel B shall be less than -50 dB at the operating frequency, 10 kHz. For example, when B's output is 1 V, A's output due to crosstalk induced by B, will be at least 50 dB down which, of course, would be less than 3.16 mV.
2. The crosstalk to channel A from channel C shall be less than -30 dB.

Proposed Circuit Board Layout (Fig. 3.3):

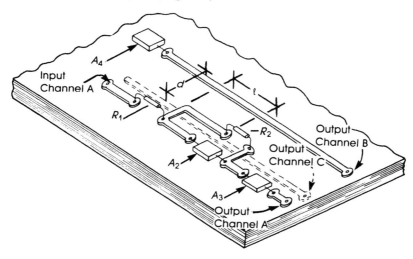

Figure 3.3 This circuit board contains 10 layers. All components are mounted on the top layer.

Circuit Board Cross Section (Fig. 3.4):

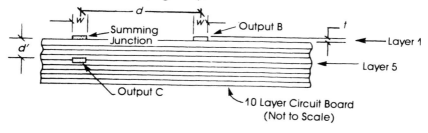

Figure 3.4 The board's cross section shows the conductors of interest.

Analysis for Channels A and B:

We will first consider the crosstalk between channels A and B because it is the more stringent of the two requirements. Using the circuit parameters shown in Fig. 3.2 and Eq. (10), we can compute the maximum allowable crosstalk capacitance, C_{ct}:

$$C_{ct}(\max) = \frac{10^{12}}{2\pi f R_1 G} \times \log^{-1}\left(\frac{K_{c(AB)}}{20}\right) \text{pF} \qquad (12)$$

$$= \frac{0.00316 \times 10^{12}}{2\pi \times 10^4 \times 2 \times 10^3 \times 2 \times 10^2} \text{ pF}$$
$$= 0.126 \text{ pF} \tag{13}$$

The proposed layout has these dimensions:

$$l = 1.5" \qquad w = 0.010"$$
$$d = 0.5" \qquad t = 0.0028"$$
$$h = 10 \times 0.007" = 0.070"$$
$$d/h = 0.5"/0.070" = 7.14$$
$$\varepsilon_r = 4.5$$

To determine the capacitance between channel B's output and the summing junction connection land in channel A, we will use Formula Set C-5. Because $d/h = 7.14$, we will use $\varepsilon_{r(\text{eff})} = (1 + \varepsilon_r)/2 = 2.75$ because this predicts a larger value of capacitance. From Formula Set C-5,

$$\frac{C}{l} \approx \frac{0.71\varepsilon_{r(\text{eff})}}{\ln\left(\dfrac{\pi d}{w + t}\right)} \text{ pF/in} \tag{14}$$

$$C \approx \frac{0.71 \times 2.75 \times 1.5"}{\ln\left(\dfrac{\pi \times 0.5"}{0.010" + 0.0028"}\right)} \text{ pF}$$

$$= 0.61 \text{ pF} \tag{15}$$

This value is larger than the maximum permitted. We can move the channel B output land farther away to get a smaller value of capacitance. Trying $d = 4"$ and using $\varepsilon_{r(\text{eff})} = 1$ (because $d/h = 4"/0.070" = 57 \gg 1$), we get

$$\frac{C}{l} \approx \frac{0.71}{\ln\left(\dfrac{\pi d}{w + t}\right)} \text{ pF/in} \tag{16}$$

$$C \approx \frac{0.71 \times 1 \times 1.5''}{\ln\left(\dfrac{\pi \times 4''}{0.010'' + 0.0028''}\right)} \text{pF}$$

$$= 0.15 \text{ pF} \qquad (17)$$

This value is still too large. This result illustrates the logarithmic nature of the capacitance magnitude with distance. We note that of the (0.61 pF − 0.15 pF) = 0.46 pF reduction, only 0.07 pF is strictly due to distance. The other 0.39 pF results from the dielectric contribution of the circuit board. (Note that end fringing has been neglected in our discussion.)

A better solution is to use a ground plane as discussed in Crosstalk Analysis CTC-2.

Analysis for Channels A and C:

The output of channel C lies directly below channel A's summing junction and is separated by four 0.007" layers of dielectric. As above, we can compute the maximum value of the crosstalk capacitance from Eq. (12). In this case, $K_{c(AC)}$ is −30 dB.

$$C_{ct}(\max) \leq \frac{10^{12}}{2\pi f R_1 G} \times \log^{-1}\left(\frac{K_{c(AC)}}{20}\right) \text{pF}$$

$$= \frac{0.0316 \times 10^{12}}{2\pi \times 10^4 \times 2 \times 10^3 \times 2 \times 10^2} \text{pF}$$

$$= 1.26 \text{ pF} \qquad (18)$$

Formula Set C-4 Eq. (2b) gives the crosstalk capacitance as

$$\frac{C}{l} = 0.225 \varepsilon_r K_{C1}\left(\frac{w}{d}\right) \text{pF/in} \qquad (19)$$

For channels A and C, the proposed layout has these dimensions:

$l = 1.5''$ $\qquad\qquad w = 0.010''$

$d' = 4 \times 0.007'' = 0.028''$ $\qquad t = 0.0028''$

$\varepsilon_r = 4.5$

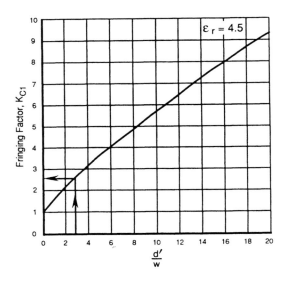

Figure 3.5 The fringing factor, K_{C1}, is 2.6 for $d/w = 0.028''/0.010'' = 2.8$.

Thus,

$$C = 0.225\varepsilon_r K_{C1}\left(\frac{w}{d'}\right) \times l \text{ pF} \tag{20}$$

$$= 0.225 \times 4.5 \times 2.6 \times \left(\frac{0.010''}{0.028''}\right) \times 1.5'' \text{ pF}$$

$$= 1.41 \text{ pF} \tag{21}$$

Here, the capacitance is slightly larger than that allowed. Channel B forced the use of a ground plane and this will make the mutual capacitance between channels A and C *very much* smaller than 1.41 pF.

Commentary and Conclusions:

1. As can be seen from examination of Eq. (6), the crosstalk level, $|E_{oa}/E_{ob}|$, is directly proportional to the operating frequency, f, the victim channel input resistance and gain, R_1 and G, and the crosstalk capacitance C_{ct}.

$$\frac{E_{oa}}{E_{ob}} = -j2\pi f R_1 C_{ct} G$$

Lowering any of these values will reduce the crosstalk.
2. EXP C-4, EXP C-5A, and EXP C-5B describe actual experiments and show test data for configurations similar to those described. These experiments show excellent correlation between predicted values and empirical results, and discuss second-order effects and other facets of practical circuit design.
3. As noted above and described in Crosstalk Analysis CTC-2, the addition of a ground plane greatly reduces crosstalk due to capacitive coupling.

3.2.2 Capacitive Coupling to Summing Junctions (with Ground Plane), CTC-2

In Crosstalk Analysis CTC-1, we saw that the crosstalk requirements could not be met because the crosstalk capacitance was too large between channels A and B. This analysis shows that a ground plane added directly below channels A and B reduces the crosstalk to well below the specified level.

Analysis:

In CTC-1, we determined the maximum value of crosstalk capacitance to achieve -50 dB between channels B and A as follows:

$$C_{ct}(\text{max}) = \frac{10^{12}}{2\pi f R_1 G} \times \log^{-1}\left(\frac{K_{c(AB)}}{20}\right) \text{pF} \tag{1}$$

$$= \frac{0.00316 \times 10^{12}}{2\pi \times 10^4 \times 2 \times 10^3 \times 2 \times 10^2} \text{pF}$$

$$= 0.126 \text{ pF} \tag{2}$$

We will now make layer 2 a ground plane as shown in Fig. 3.6.
The proposed layout has these dimensions:

$$l = 1.5" \qquad w = 0.010"$$
$$d = 0.5" \qquad t = 0.0028"$$

We can determine the new capacitance between channel B's output and the summing junction connection land in channel A from Formula Set C-7:

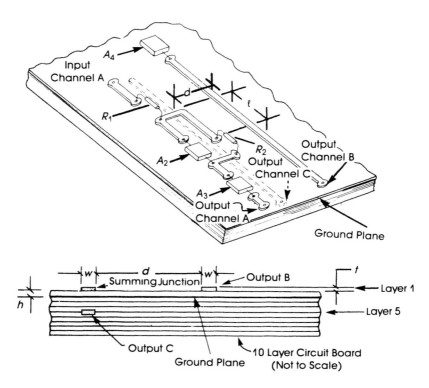

Figure 3.6 This is the same structure shown in CTC-1, Fig. 3.3, but with a ground plane added to layer 2. This not only reduces the crosstalk between channels B and A, but also effectively shields channel A from the output of channel C.

$$\frac{C_m}{l} = \frac{C_{mct}}{l} \approx 0.07\varepsilon_r K_{L1} K_{C1} \left(\frac{w}{d}\right)^2 \text{ pF/in,} \quad \text{for} \quad \frac{2h}{d} < 0.3 \qquad (3)$$

The ratio $2h/d = 2 \times 0.007''/0.5'' = 0.028$, which is much smaller than 0.3. For the single layer $h = 0.007''$, and therefore $2h/w = 2 \times 0.007''/0.010'' = 1.4$. The fringing factors, K_{L1} and K_{C1} are, from Fig. 2.30, 2.4 and 1.8 respectively. The mutual crosstalk capacitance, C_{mct}, is then

$$C_{mct} = 0.07 \times 4.5 \times 2.4 \times 1.8 \times \left(\frac{0.010''}{0.5''}\right)^2 \times 1.5'' \text{ pF}$$

$$= 0.00082 \text{ pF} \qquad (4)$$

Commentary and Conclusions:

1. Comparison of the results of this analysis and that of CTC-1 shows that the addition of the ground plane reduces crosstalk by 0.61 pF/0.00082 pF = 747 or 57 dB. The magnitude of the new crosstalk theoretically would be

$$K_c = 20 \log \left| \frac{E_{oa}}{E_{ob}} \right|$$

$$= 20 \log (2\pi f R_1 C_{ct} G) \text{ dB}$$

$$= 20 \log (2\pi \times 10^4 \times 2 \times 10^3 \times 8.2 \times 10^{-16} \times 2 \times 10^2)$$

$$= -94 \text{ dB} \qquad (5)$$

compared with the -50 dB requirement. Note that -94 dB/1 V is only 21 μV crosstalk to channel A when the output of channel B is 1 V. At this low level, other factors contributing to crosstalk need to be considered.

2. In order to achieve low crosstalk values, very close attention must be paid to summing-junction length, l, distance, d, and height, h.

3.3 INDUCTIVE COUPLING BETWEEN CIRCUITS, CTL-1

In addition to capacitive coupling discussed in Crosstalk Analyses CTC-1 and CTC-2 crosstalk between circuits can occur due to inductive coupling. In this analysis, we will consider inductive coupling on printed wiring boards. The relative magnitudes between capacitive and inductive crosstalk are also investigated.

Analysis:

Figure 3.7 is a schematic representing inductive coupling between the output of channel B and the input of channel A. Assume that the circuit is on a single PWB as shown in Fig. 3.8. Figure 3.9 shows the equivalent circuit.

The voltage, E'_{ia} is the crosstalk voltage induced in the input of channel A due to the output current of channel B. From the definition of mutual inductance, the value of this voltage is given by

$$E'_{ia} = L_m \frac{di_b(t)}{dt} \qquad (1)$$

The current flowing in loop b is

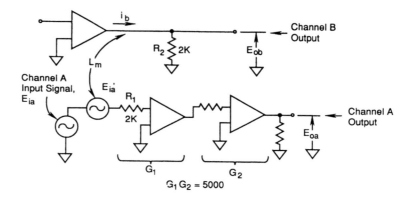

Figure 3.7 A time-varying current, i_b, in channel B induces a voltage E'_{ia} in the input of channel A.

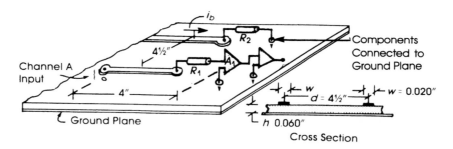

Figure 3.8 A ground plane is located 1 layer below the circuit lands. The return for current i_b is via the ground plane.

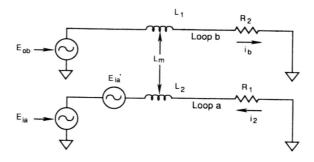

Figure 3.9 Mutual inductance, L_m, links the two circuits.

$$i_b(t) = I_{b\max} \sin \omega t \text{ A} \tag{2}$$

and

$$\frac{di_b(t)}{dt} = \omega I_{b\max} \cos \omega t = 2\pi f I_{b\max} \cos\omega t \tag{3}$$

The rms magnitude of the induced voltage in loop *a* is then

$$|E'_{ia}| = 2\pi f L_m i_{b\text{rms}} \tag{4}$$

$$= 2\pi f L_m \frac{E_{ob}}{R_2} \tag{5}$$

The output of channel A is

$$E_{oa} = |E'_{ia}| \times G$$

$$= 2\pi f L_m \frac{E_{ob}}{R_2} \times G \tag{6}$$

The inductive crosstalk, K_l, is then

$$K_l = 20 \log \left|\frac{E_{oa}}{E_{ob}}\right| = 20 \log \left(2\pi f L_m \frac{G}{R_2}\right) \tag{7}$$

Formula Set L-7 gives the mutual inductance, L_m, as

$$\frac{L_m}{l} = 0.01 \left(\frac{h}{d}\right)^2 \ \mu\text{H/in} \tag{8}$$

Substituting the values $h = 0.060''$, $d = 4.5''$ and $l = 4''$ in Eq. (8) yields

$$L_m = 0.01 \left(\frac{0.060''}{4.5''}\right)^2 \times 4'' \ \mu\text{H}$$

$$= 7.1 \times 10^{-6} \ \mu\text{H} \tag{9}$$

For $f = 50$ kHz, Eq. (7) gives the crosstalk as

$$K_l = 20 \log \left|\frac{E_{oa}}{E_{ob}}\right|$$

$$= 20 \log \left(2\pi \times 50 \times 10^3 \times 7.1 \times 10^{-12} \times \frac{5 \times 10^3}{2 \times 10^3}\right)$$

$$= -105 \text{ dB} \qquad (10)$$

Commentary and Conclusions:

1. We recommend that pertinent circuit values be estimated at the start of a crosstalk investigation and the analysis be directed accordingly.
2. Using Crosstalk Analyses CTC-1 and CTC-2, we can compute the capacitive crosstalk expected for the circuit shown in Fig. 3.7. Equation (7) from CTC-1 is

$$K_c = 20 \log \left|\frac{E_{oa}}{E_{ob}}\right|$$

$$= 20 \log (2\pi f R_1 C_{mct} G) \text{ dB} \qquad (11)$$

(Because we are using a ground plane, C_{ct} becomes C_{mct}.) From CTC-2, Eq. (3) the mutual crosstalk capacitance per unit length, C_{mct}/l is given by

$$\frac{C_{mct}}{l} = 0.07 \varepsilon_r K_{L1} K_{C1} \left(\frac{w}{d}\right)^2 \text{ pF/in}, \quad \text{for} \quad \frac{2h}{d} < 0.3 \qquad (12)$$

In this case, the ratio $2h/d = 2 \times 0.060''/4.5'' = 0.027$, which is much smaller than 0.3. Because $h = 0.060''$, making $2h/w = 2 \times 0.060''/0.020'' = 6$, the fringing factors K_{L1} and K_{C1} are, from Fig. 2.30, 5.9 and 4.00 respectively. The mutual crosstalk capacitance is then

$$C_{mct} = 0.07 \times 4.5 \times 5.9 \times 4.0 \times \left(\frac{0.020''}{4.5''}\right)^2 \times 4'' \text{ pF}$$

$$= 5.9 \times 10^{-4} \text{ pF} \qquad (13)$$

Putting the values in Eq. (11) we get:

$$K_c = 20 \log \left|\frac{E_{oa}}{E_{ob}}\right|$$

$$= 20 \log (2\pi \times 50 \times 10^3 \times 2 \times 10^3 \times 5.9 \times 10^{-16} \times 5 \times 10^3)$$
$$= -54.6 \text{ dB} \tag{14}$$

Equation (10) found the inductive crosstalk to be $K_I = -105$ dB. Comparing the difference:

$$K_C - K_I = -54.6 \text{ dB} - (-105 \text{ dB})$$
$$= 50.4 \text{ dB} \tag{15}$$

Thus, the capacitive crosstalk is 50.4 dB greater than the inductive crosstalk in this example.

3. The results found above can be expressed in more general terms. Equations (11) and (7) can be combined to get the ratio between capacitive and inductive crosstalk:

$$\log^{-1}\left(\frac{K_C - K_I}{20}\right) = \frac{2\pi f R_1 C_{mct} G}{2\pi f L_m \dfrac{G}{R_2}}$$

$$= \frac{R_1 R_2 C_{mct}}{L_m} \tag{16}$$

Formula Set C-7 gives the following ratio for lands equal heights above the ground plane:

$$\frac{C_m}{l} = \frac{\left(\dfrac{L_m}{l}\right)\left(\dfrac{C_1}{l}\right)}{\left(\dfrac{L_1}{l}\right)} \tag{17}$$

From Formula Sets C-6 and L-6,

$$\frac{C}{l} = \varepsilon_r \varepsilon_o K_{C1} \left(\frac{w}{h}\right) \text{ F/m} \tag{18}$$

$$\frac{L}{l} = \frac{\mu_r \mu_o}{K_{L1}} \left(\frac{h}{w}\right) \text{ H/m} \tag{19}$$

Thus,

$$\frac{C_m}{l} = \frac{\left(\dfrac{L_m}{l}\right)\left[\varepsilon_r \varepsilon_0 K_{C1}\left(\dfrac{w}{h}\right)\right]}{\left[\dfrac{\mu_r \mu_0}{K_{L1}}\left(\dfrac{h}{w}\right)\right]} \quad (20)$$

Multiplying both sides of Eq. (20) by l, letting $\mu_r = 1$, $\mu_0/\varepsilon_0 = (120\pi)^2$, and simplifying, we get

$$\frac{C_m}{L_m} = \frac{\varepsilon_r K_{C1} K_{L1}}{(120\pi)^2}\left(\frac{w}{h}\right)^2 \quad (21)$$

Formula Set Z_0-(ALL), Z_0-6 shows that the right-hand side of Eq. (21) is $1/(Z_{01})^2$, where Z_{01} is the characteristic impedance of a flat conductor over a ground plane.

Combining Eqs. (16) and (21) and generalizing by letting $(Z_{01})^2 = Z_{01}Z_{02}$, produces:

$$\log^{-1}\left(\frac{K_C - K_I}{20}\right) = \frac{R_1 R_2}{Z_{01} Z_{02}} \quad (22)$$

Equation (22) illustrates an important principle. If the product of the source impedance, R_2, and the receiver impedance, R_1, is large compared with the product of the land-to-ground-plane characteristic impedances, the crosstalk will be mostly capacitive. If the inverse is true, the crosstalk will be mostly inductive. In other words,

If $R_1 R_2 \gg Z_{01} Z_{02} \rightarrow$ capacitive crosstalk (23)

If $R_1 R_2 \ll Z_{01} Z_{02} \rightarrow$ inductive crosstalk (24)

3.4 COMMON GROUND COUPLING, CTG-1

The use of single-point signal grounds can greatly reduce crosstalk because signals are not coupled by currents flowing through ground lines shared by two or more channels. Furthermore, locating the power supplies down-stream from the low signal level amplifiers can provide significant crosstalk reductions. Experiments EXP CTG-1A, CTG-1B, and CTG-1C show empirical data validating the ideas presented in this section.

Shared (Common) Ground Bus Analysis:

Figure 3.10 represents two identical channels with a common signal and power supply ground return.

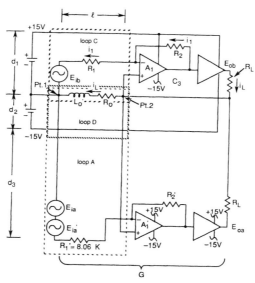

Figure 3.10 In this simplified diagram, the inputs E_{ia} and E_{ib} are tied to Pt. 1. The amplifier grounds are connected to Pt. 2. The two channels thus share the common ground impedances L_0' and R_0.

The crosstalk between Channel B and Channel A can be computed as follows:
For regular signals,

$$E_{oa} = GE_{ia} \tag{1}$$

For crosstalk signals,

$$E_{oa} = GE_{ia}' \tag{2}$$

with

$$E_{ia}' = \sqrt{R_0^2 + (X_{L_0'})^2} \times i_L \tag{3}$$

where

R_0 = resistance of shared signal-ground return, Ω
$X_{L_0'}$ = equivalent inductive reactance of shared signal-ground return, Ω
    = $2\pi f L_0'$
L_0' = equivalent self-inductance of shared signal-ground return, H

The "load" current is given by

$$i_L = \frac{E_{ob}}{R_L} \tag{4}$$

Combining Eqs. (2), (3), and (4) to determine the ground coupling crosstalk, K_g,

$$K_g = 20 \log \left| \frac{E_{oa}}{E_{ob}} \right| = 20 \log \left(\frac{G\sqrt{R_0^2 + X_{L_0'}^2}}{R_L} \right) \quad (5)$$

Example:

Determine the ground coupling crosstalk for the circuit shown in Fig. 3.10, assuming these parameters:

Loop C: $d_1 = 3.1''$, $l = 3''$

Loop D: $d_2 = 0.4''$, $l = 3''$

Channel B/A interconnections forming loop A: $d_3 = 8.2''$

Land widths, $w = 0.025''$ with thickness, $t = 0.0028''$

$G = 1051$, $R_L = 1.8 \text{ k}\Omega$

R_0 can be calculated as follows:

From Formula Set R-1,

$$R_0 = 0.394 \rho' \frac{l}{w \times t} \, \Omega \quad (6)$$

For PWB lands 0.025" wide and 0.0028" high (2 oz. copper),

$$R_0 = 0.394 \times 1.725 \times 10^{-6} \frac{3''}{0.025'' \times 0.0028''} \, \Omega$$

$$= 0.029 \, \Omega \quad (7)$$

Calculation of the equivalent inductance L_0' is complicated by several factors which are discussed in the Commentary and Conclusions, item 1. With those comments in mind, we will calculate a value for L_0' (which will be slightly overstated).

From Formula Set L-5,

$$\frac{L}{l} \approx 0.01 \ln \left(\frac{\pi d}{w + t} \right) \, \mu\text{H/in} \quad (8)$$

The total inductance of loop C is then

$$L_0(\text{loop C}) = 0.01 \ln\left(\frac{\pi \times 3.1''}{0.025'' + 0.0028''}\right) \times 3'' \ \mu H$$

$$= 0.18 \ \mu H \qquad (9)$$

and for loop D:

$$L_0(\text{loop D}) = 0.01 \ln\left(\frac{\pi \times 0.4''}{0.025'' + 0.0028''}\right) \times 3'' \ \mu H$$

$$= 0.11 \ \mu H \qquad (10)$$

For purposes of this calculation, we can consider that the equivalent inductance, L_0', is one-half the average of the two inductances found in Eqs. (9) and (10). Thus,

$$L_0' = \frac{1}{2} \times \frac{1}{2} \times [L_0(\text{loop C}) + L_0(\text{loop D})]$$

$$= \frac{1}{4} \times (0.18 + 0.11)$$

$$= 0.07 \ \text{mH} \qquad (11)$$

At $f = 50$ kHz,

$$X_{L_0'} = 2\pi f L_0'$$

$$= 2\pi \times 50 \times 10^3 \times 0.07 \times 10^{-6}$$

$$= 0.022 \ \Omega \qquad (12)$$

Substituting the values in Eq. (5),

$$K_g = 20 \log\left(\frac{G\sqrt{R_0^2 + (X_{L_0'})^2}}{R_L}\right)$$

$$= 20 \log\left[\frac{1.051 \times 10^3 \times \sqrt{(0.029)^2 + (0.022)^2}}{1.8 \times 10^3}\right]$$

$$= -33.5 \ \text{dB} \qquad (13)$$

Single-Point Ground Analysis:

In Fig. 3.11 Channels A and B are provided with separate ground buses and all of the grounds are connected at a single point.

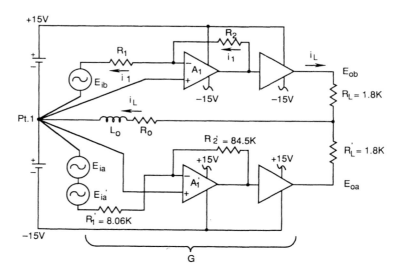

Figure 3.11 The channel inputs and grounds are connected to Point 1.

Theoretically, there should be no common ground crosstalk between channels B and A. Measurements made in EXP CTG-1b show 14 dB crosstalk level improvement to −42.4 dB. Lower levels were not achieved because of the amplifier noise floor. In applying single-point grounds, precautions such as using common-mode isolation amplifiers should be made to prevent reconnecting the grounds at some downstream point (such as a multiplexer).

Downstream Power Supply Location:

Another solution is offered. If the power supplies are moved to the high-level amplifier location shown in Fig. 3.12, the current i_L will be eliminated from the input amplifiers A_1 and A_1' ground bus and replaced with i_1. Additionally, the gain term, G, is no longer a factor.

The crosstalk is calculated in a similar manner as above:

$$E_{oa} = G E_{ia}' \tag{14}$$

In this case,

$$E_{ia}' = \sqrt{R_0^2 + (X_{L_0})^2} \times i_1 \tag{15}$$

where

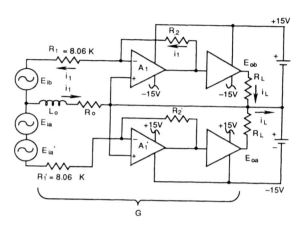

Figure 3.12 The ±15 V power supplies are moved to the load end of the circuit.

$$i_1 = \frac{E_{ib}}{R_1} \tag{16}$$

and

$$E_{ob} = G E_{ib} \tag{17}$$

Combining Eqs. (14), (15), (16), and (17) to determine the new ground coupling crosstalk, K'_g. Thus,

$$K'_g = 20 \log \left| \frac{E_{oa}}{E_{ob}} \right| = 20 \log \left(\frac{\sqrt{R_0^2 + (X_{L_0})^2}}{R_1} \right) \tag{18}$$

Using the values from Eqs. (7) and (10) with $R_1 = 8.06$ kΩ,

$$K'_g = 20 \log \left[\frac{\sqrt{(0.029)^2 + (0.022)^2}}{8.06 \times 10^3} \right]$$

$$= -106.9 \text{ dB} \tag{19}$$

Equation (19) predicts a very low crosstalk level. However, EXP CTG-1C shows measured results of -38 dB. This discrepancy illustrates two important points:

1. While moving the power supplies downstream resulted in a 9.2 dB improvement, the predicted value was not achieved, by a considerable margin, due to the noise floor.
2. Achieving low crosstalk levels requires close attention to *all* pertinent circuit layout aspects. In this case, power supply, inductive and other coupling produced the dominant crosstalk.

Commentary and Conclusions:

1. As noted, calculation of X'_{L0} is complicated by two factors:
   - The ± 15 V sources alternately feed the current to the load, i.e., the $+15$ V supplies the positive half-cycle, the -15 V the negative half-cycle. The ± 15 V return contains both half-cycles of the current.
   - The magnetic field caused by these currents induces a voltage in loop A. This voltage is proportional to the time rate of change of the flux, or

$$E_{iA} = \frac{d\Psi_A}{dt} \tag{20}$$

where Ψ_A = total flux in loop A due to current contained in circuit B. Because self-inductance is defined as

$$L = \frac{\Psi}{i} \tag{21}$$

then

$$\frac{d\Psi_A}{dt} = L\frac{di}{dt} \tag{22}$$

and, therefore,

$$E'_{iA} = E'_{ia} = L'_0 \frac{di}{dt} \tag{23}$$

The land patterns are shown in Fig. 3.13.
The following observations are made with regard to Fig. 3.13:

1. During the positive half-cycle, current flowing in the $+15$ V and ± 15 V RTN conductors generates a magnetic flux field, which is similar in concept to that shown in Appendix 1, Fig. A1.4, although the conductor-dimensions-to-separation-distance ratios are quite different between the two cases.
2. The flux lines, one of which is symbolized by Ψ_1, due to the $+15$ V conductor, are entirely to the right of the dashed line plane and therefore do not link loop A.
3. On the contrary, most of the ± 15 V RTN flux lines, represented by Ψ_2, link loop A.
4. Current flowing in the ± 15 V RTN and the -15 V conductors during the negative half-cycle generates flux represented by Ψ_3 and Ψ_4.

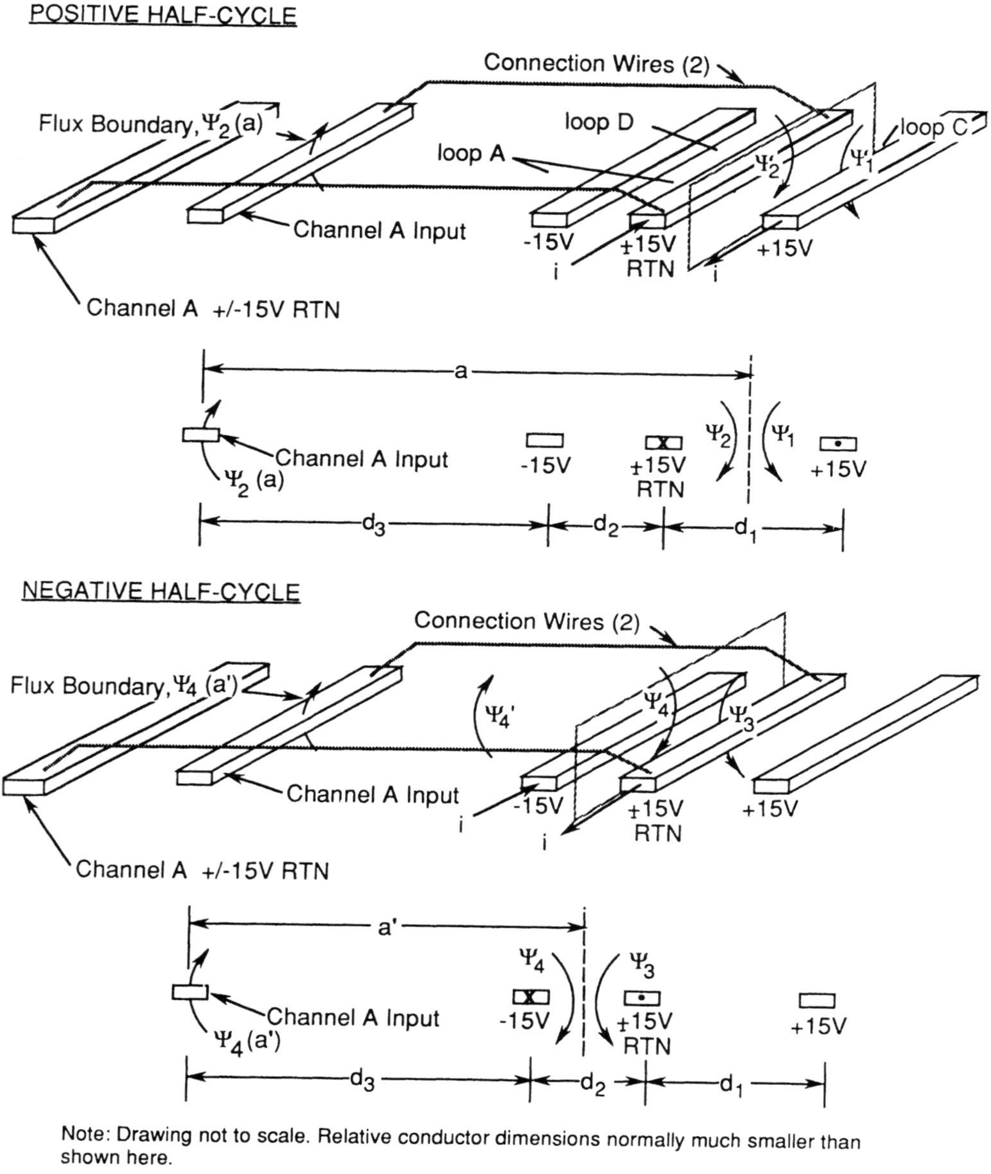

POSITIVE HALF-CYCLE

NEGATIVE HALF-CYCLE

Note: Drawing not to scale. Relative conductor dimensions normally much smaller than shown here.

Figure 3.13 The self-inductance, L_0', is determined by the flux linkages between circuits A and B. These are sketched for the two half-cycles. For the both half-cycles, the flux linking loop A is due almost entirely to the current in the ± 15 V return.

5. As Fig. 3.13 suggests, all of the Ψ_3 flux lines link loop A.
6. While the flux, represented by Ψ_4, produced by current in the -15 V conductor, also completely links loop A, but almost all of the flux is canceled by flux continuation lines represented by Ψ'_4. Thus, the net flux in loop A due to the -15 V conductor current is almost zero.

The following analysis quantifies the flux linkages for the conductors. Appendix 1 gives the ratio, α, between the values of the flux boundary $\Psi(a)$, measured in a units from the x-axis origin, to that at the conductor as

$$\alpha \approx \frac{\ln\left(\dfrac{2a + d}{2a - d}\right)}{\ln\left(\dfrac{\pi d}{w + t}\right)} \tag{24}$$

For the positive half-cycle ± 15 V RTN case, $d = 3.1''$ and $a = 1.55'' + 0.4'' + 8.2'' = 10.15''$. Substituting these values in Eq. (24), we get the ratio of the value of the flux boundary $\Psi_2(a)$, at the input to channel A, to the flux boundary at the ± 15 V RTN conductor as

$$\alpha(+15 \text{ V}) \approx \frac{\ln\left[\dfrac{(2 \times 10.15'') + 3.1''}{(2 \times 10.15'') - 3.1''}\right]}{\ln\left(\dfrac{\pi \times 3.1''}{0.025'' + 0.0028''}\right)}$$

$$= 0.05 \tag{25}$$

The flux linking loop A is $1 - \alpha$. Therefore, almost all of the flux generated by the ± 15 V RTN current links loop A.

For the negative half-cycle, we saw previously that all of the flux associated with the ± 15 V RTN conductor links loop A. We are therefore concerned with the flux produced by the current in the -15 V conductor that *does not* link loop A, the one beyond flux boundary $\Psi_4(a)$.

For the negative half-cycle then, $d = 0.4''$ and $a' = 0.2'' + 8.2'' = 8.4''$.

$$\alpha'(-15 \text{ V}) \approx \frac{\ln\left[\dfrac{(2 \times 8.4'') + 0.4''}{(2 \times 8.4'') - 0.4''}\right]}{\ln\left(\dfrac{\pi \times 0.4''}{0.025'' + 0.0028''}\right)}$$

$$= 0.01 \tag{26}$$

Accordingly, only 1% of the total flux produced by the -15 V conductor current, during the negative half-cycle, links loop A.

We have therefore seen that, for the geometry considered in this case, almost all of the flux generated by the ± 15 RTN conductor current links the victim circuit.

We should mention that the current flowing in the ± 15 V RTN would be sinusoidal (except for minor distortion caused by the amplifiers). The currents flowing in the $+15$ V and -15 V conductors are half-wave-rectified sine waves. These latter currents contain the fundamental frequency and its harmonics. The amplitude of the fundamental is half that of the amplitude flowing in ± 15 V RTN.

3.5 CIRCUIT CROSSTALK DUE TO POWER SUPPLIES, CTPS-1

Power Supplies can indirectly produce crosstalk because of nonzero output impedances and improper ground connections. This analysis calculates the voltage caused by power supply bus resistances and inductances and suggests a method for reducing the crosstalk.

Analysis:

Consider Fig. 3.14 which is a circuit diagram showing three ideal power supplies: $+15$ V, -15 V, and $+5$ V connected to a single-point ground. Figure 3.15 illustrates a realistic circuit configuration.

Crosstalk voltages will appear at the loads of each supply due to currents in the other two. These voltages are caused by the currents through the common bus impedances L_C and R_C.

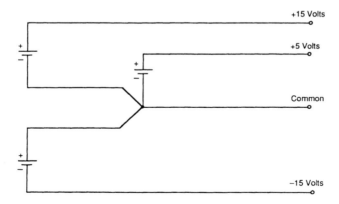

Figure 3.14 These ideal power sources produce constant direct current without noise through zero-impedance buses.

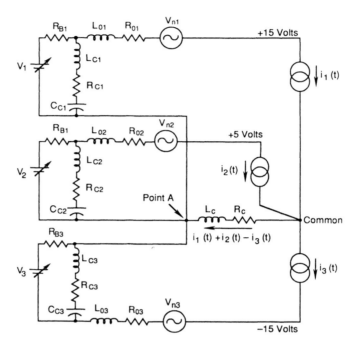

Figure 3.15 Power supply noise sources and bus impedances are added to the ideal sources shown in Fig. 3.14.

We can calculate the magnitude of the crosstalk voltage for in-phase currents of the same frequency as follows.

$$V_{ct} = (i_1 + i_2 - i_3) \times Z_{ct} \text{ V} \tag{1}$$

$$= (i_1 + i_2 - i_3) \times \sqrt{(R_C)^2 + (2\pi f L_C)^2} \text{ V} \tag{2}$$

Example:

Calculate the crosstalk voltage on the -15 V supply due to a 4 A rms load at $f = 50$ kHz on the $+5$ V supply for the power distribution system shown in Fig. 3.16.

Calculating the inductance from Formula Set L-1:

$$2L_C \approx 0.01 \ln\left(\frac{d}{r}\right) \times l \text{ } \mu\text{H} \tag{3}$$

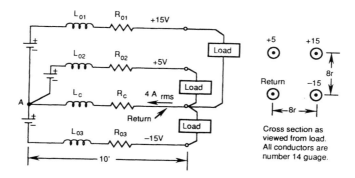

Figure 3.16 The power supplies deliver current through #14 gauge conductors separated by a distance of four wire diameters with a one-way distance of ten feet.

$$= 0.01 \ln\left(\frac{8r}{r}\right) \times 120'' \ \mu\text{H} \qquad (4)$$

$$= 2.5 \ \mu\text{H}$$

From wire tables #14 wire has a dc resistance of 0.0025 Ω/ft and a radius $r = 0.032''$. From Ref. [1], the ac resistance is given by

$$R_{ac} = [(0.096 \times 2r \times \sqrt{f}) + 0.26]R_{dc}$$
$$= [(0.096 \times 0.064 \times \sqrt{5 \times 10^4}) + 0.26] \times 0.0025 \times 10'$$
$$= 1.63 \times 0.0025 \times 10'$$
$$= 0.04 \ \Omega \qquad (5)$$

Calculating $X_C = 2\pi \times 5 \times 10^4 \times 1.25 \times 10^{-6} = 0.39 \ \Omega$, and then Z_C, we get

$$Z_C = \sqrt{(R_C)^2 + (2\pi f L_C)^2}$$
$$= \sqrt{(0.04)^2 + (0.39)^2}$$
$$\approx 0.39 \ \Omega \qquad (6)$$

With $i_2 = 4$ A, we can compute the crosstalk voltages using Eq. (1):

$$V_{ct} = i_1 \times Z_{ct} \ \text{V}$$
$$= 4 \times 0.39$$
$$= 1.56 \ \text{V} \qquad (7)$$

Thus, 1.56 V will be added, as "noise" to the −15 V supply voltage (and to the +15 V as well).

Commentary and Conclusions:

1. Please note that no useful purpose is achieved by running separate returns for each supply voltage as long as they are tied together at point A. The solution lies in providing single point grounds located *at the point of use,* as shown in Fig. 3.11, Crosstalk Analysis CTG-1.
2. The preceding analysis is based on the voltage drop occurring on the common ground between the supplies. Other effects such as capacitance and mutual inductance between the various conductors were not included.

REFERENCES

1. Ott, H.W., *Noise Reduction Techniques in Electronic Systems,* New York, John Wiley and Sons 1976, p. 129.

Chapter 4
Discrete Components

4.1 INTRODUCTION

Capacitors and inductors play key decoupling and filtering roles in electronic circuit design. In many instances, however, the addition of these parts affords little or no reduction in noise or crosstalk levels.

This chapter presents equivalent circuits for realizable components, the resulting impedance *versus* frequency curves and measured data for some typical parts.

4.2 CHARACTERISTICS OF COMMONLY USED CAPACITORS, DC-1

The capacitor is an important discrete circuit component use for filtering, providing a local low-impedance energy source and noise decoupling between circuits. The examples show that capacitors are, in reality, series *LRC* circuits and have "cut-off" frequencies beyond which they appear as inductors.

Introduction:

The ideal capacitor is shown in Fig. 4.1a. It has no series inductance, equivalent series resistance (ESR), or parallel leakage resistance. Figure 4.1b shows the equivalent circuit for realizable capacitors.

Discussion:

Figures 4.2 and 4.3 show the impedance as a function of frequency for the equivalent circuit (Fig. 4.1b).

Aluminum and tantalum electrolytic capacitors exhibit impedance *versus* frequency characteristics as shown by the smooth curve of Fig. 4.2. Following the

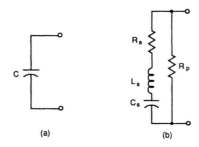

Figure 4.1 An ideal capacitor and the equivalent circuit for realizable capacitors.

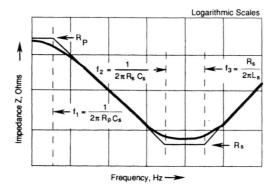

Figure 4.2 Capacitor impedance *versus* frequency when $Q < 1$.

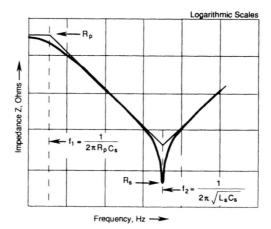

Figure 4.3 Capacitor impedance *versus* frequency when $Q > 1$.

asymptotic approximations to the real curve, we see that the impedance, Z, is flat with frequency, and is equal to the parallel resistance, R_p. At f_1, the impedance falls with a -6 dB/octave slope. The equivalent series resistance, R_s, causes the relatively flat characteristic between f_2 and f_3. Beyond f_3, the series inductance becomes the dominant factor producing a slope of $+6$ dB/octave. One note of caution—capacitors, especially electrolytics, can have frequency-dependent nonlinearities and parasitic elements in addition to those shown in Fig. 4.1b.

Ceramic, mica, and other types of high frequency capacitors generally have a Q-factor greater than 1. This produces a pronounced series resonance at the resonant frequency, $\omega_0 = 2\pi f_0$ with the impedance first decreasing and then increasing very rapidly, as shown near f_2 in Fig. 4.3.

Example 1:

Figure 4.4 shows the plotted test data for a solid tantalum electrolytic capacitor. From this data, determine the capacitance, ESR, and series inductance.

Using the data, we can calculate the equivalent circuit values:

Capacitance, C_s:

Selecting $f = 100$ Hz, $|Z| = 36$ Ω, and solving for C_s, we get

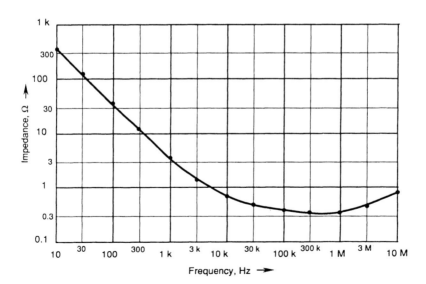

Figure 4.4 Z *versus* frequency for a 47 μF $\pm$ 10 percent, 35 V solid tantalum electrolytic capacitor.

$$C_s \approx \frac{1}{\omega |Z|} = \frac{1}{2\pi f |Z|} = \frac{1}{2\pi \times 100 \times 36} \text{ F}$$

$$= 44.2 \ \mu\text{F (within the } \pm 10 \text{ percent tolerance)} \tag{1}$$

Equivalent Series Resistance, R_s:

The "flat" portion of the curve actually slopes slightly downward showing that the ESR is frequency-dependent. Taking the middle value, we get

$$R_s = 0.47 \ \Omega \tag{2}$$

Series Inductance, L_s:

$|Z|$ at 10 MHz is 0.860 Ω. Solving for L_s, we get

$$L_s = \frac{|Z|}{\omega} = \frac{|Z|}{2\pi f} = \frac{0.860}{2\pi \times 10} \ \mu\text{H}$$

$$= 0.014 \ \mu\text{H} \tag{3}$$

We can compare this measured result with an order-of-magnitude value calculated from the physical geometry shown in Fig. 4.5. We should not, however, expect close correlation because of the small inductance values, capacitor body effects, *et cetera*.

Formula Set L-1 gives the inductance of the #22 gauge vertical leads as $L_{s-1} = 0.016 \ \mu$H. Using Formula Set L-2 and calculating two separate inductances, one for the #22 gauge leads ($r = 0.0125''$) and the other for the slug or case ($r = 0.14''$), with each $\approx 0.5''$ long,

$$\frac{L}{l} \approx 0.005 \ \ln\left(\frac{2h}{r}\right) \ \mu\text{H/in} \tag{4}$$

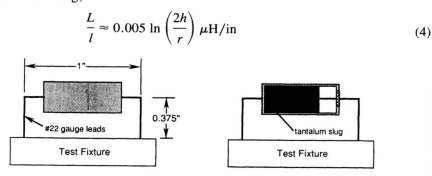

Figure 4.5 The capacitor mounting, in part, determines the cut-off frequency.

$$L_{s-2} \approx 0.005 \times 0.5'' \times \ln\left(\frac{2 \times 0.375''}{0.0125''}\right) \mu H$$

$$= 0.010 \; \mu H \text{ (for the #22 gauge leads)} \tag{5}$$

$$L_{s-3} \approx 0.005 \times 0.5'' \times \ln\left(\frac{2 \times 0.375''}{0.14''}\right) \mu H$$

$$= 0.004 \; \mu H \text{ (for the 0.28'' diameter case)} \tag{6}$$

$$L_s = L_{s-1} + L_{s-2} + L_{s-3}$$

$$= 0.030 \; \mu H \text{ versus } 0.014 \; \mu H \text{ measured} \tag{7}$$

Parallel Resistance, R_p:

Since R_p is in the MΩ range, its effect is not indicated by Fig. 4.4.

Example 2:

Figure 4.6 shows the plotted test data for a 1 $\mu F \pm$ 10% 50 V ceramic capacitor. Determine the capacitance, ESR, and series inductance from the measured data.

Following Example 1, we can calculate the equivalent circuit values:

Capacitance, C_s:

Selecting $f = 100$ kHz, $|Z| = 1.67 \; \Omega$, and solving for C_s, we get

$$C_s \approx \frac{1}{\omega|Z|} = \frac{1}{2\pi f |Z|}$$

$$= \frac{1}{2\pi \times 10^5 \times 1.67} F$$

$$= 0.953 \; \mu F \text{ (within the } \pm 10\% \text{ tolerance)} \tag{8}$$

Equivalent Series Resistance, R_s:

In this case, Q is greater than 1 and the series resonance forms a notch in the impedance curve with a minimum value at $f = 2$ MHz. The ESR is this minimum impedance value. Thus,

$$R_s = 0.014 \; \Omega \tag{9}$$

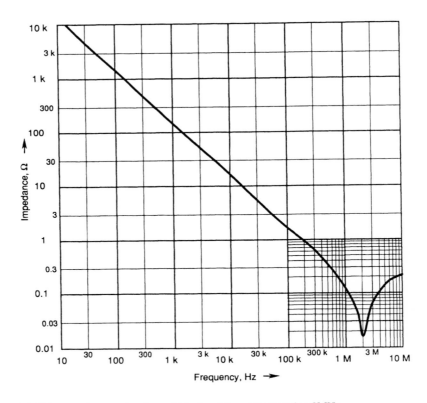

Figure 4.6 This ceramic capacitor has a $Q > 1$ and is resonant at $f = 2$ MHz.

Series Inductance, L_s:

The series resonant frequency is 2 MHz. We can solve for the series inductance from the relationship:

$$\omega_0 = \frac{1}{\sqrt{LC}}$$

$$L = \frac{1}{C(\omega_0)^2} \tag{10}$$

$$= \frac{1}{0.953 \times 10^{-6} \times (2\pi)^2 \times (2 \times 10^6)^2}$$

$$= 6.64 \text{ nH} \tag{11}$$

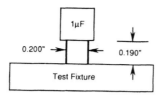

Figure 4.7 Measurement setup.

As we did with the 47 μF capacitor, we can compare the result derived from the measured data with that calculated from the physical geometry shown in Fig. 4.7.

Formula Set L-12 is used in this case because the displacement current flows up one of the #22 gauge leads, to the plates of the capacitor, and down the other. Since the plates are essentially a short circuit at this frequency, we have a rectangular loop $0.200'' \times 0.190''$. This has an area equivalent to a circular loop with $R = 0.1''$. From L-12,

$$L = 0.032\, R\left[\left(\ln \frac{8R}{r}\right) - 2\right] \mu H$$

$$= 0.032 \times 0.1'' \times \left[\left(\ln \frac{8 \times 0.1''}{.0125''}\right) - 2\right] \mu H$$

$$= 6.9 \text{ nH, compared with 6.64 nH found by Eq. (10)}. \qquad (12)$$

Parallel Resistance, R_p:

As with the 47 μF capacitor, the parallel resistance is in the MΩ range, and hence does not appear in the measured data.

Commentary and Conclusions:

1. The examples show that the series inductance is caused by the capacitor and lead geometry. Thus, when using capacitors for decoupling or filtering purposes, to minimize lead lengths is important, and thereby we avoid adding extra inductance in series with the capacitor. A good practice is to form two single-point connections consisting of the input land, the output land, and capacitor lead (for both sides of the capacitor).

4.3 CHARACTERISTICS OF COMMONLY USED INDUCTORS, DL-1

The inductor is a discrete circuit component used for filtering, tuned circuits, and other applications. The example shows that inductors are, in reality, parallel *LRC* circuits and have cut-off frequencies above which they appear as capacitors. In this way, the inductors are somewhat analogous to the capacitor behavior we have seen in Discrete Component DC-1.

Introduction:

The ideal inductor is shown in Fig. 4.8a. It has no series resistance or parallel capacitance. Figure 4.8b shows the equivalent circuit for realizable inductors.

Discussion:

Figure 4.9 shows the impedance as a function of frequency for the equivalent circuit (Fig. 4.8b).

Example 1:

From the plotted test data for a 55 μH +15%, −5% inductor shown in Fig. 4.10, determine L_s, R_s, C_p, and R_p.

Using the test data, we can calculate the equivalent circuit values:

Inductance, L_s:

Selecting $f = 100$ kHz, $|Z| = 35$ Ω, and solving for L_s, we get

$$L_s \approx \frac{|Z|}{\omega} = \frac{|Z|}{2\pi f}$$

$$= \frac{35}{2\pi \times 10^5} \text{H}$$

$$= 55.7 \ \mu\text{H (within the +15\%, −5\% tolerance)} \tag{1}$$

Series Resistance, R_s:

The series resistance of the winding, measured at 10 Hz, was found to be 0.01 Ω *versus* 0.02 Ω specified. Skin effect causes the winding resistance to increase with frequency.

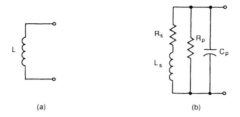

Figure 4.8 An ideal inductor (a). The physically realizable inductor (b) has additional elements that limit the performance.

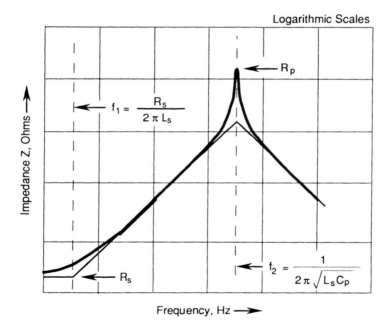

Figure 4.9 Inductor Impedance *versus* Frequency. Below f_1, the inductor impedance is primarily determined by the winding dc resistance, R_s. Above f_1, the inductance, L_s, becomes dominant and the impedance increases with a +6 dB/octave slope. At $\omega_0(f_2)$, a parallel resonance develops between L_s and C_p. Beyond f_2, the inductor becomes essentially a capacitor. As with capacitors, inductors can exhibit additional parasitic elements, especially at the higher frequencies.

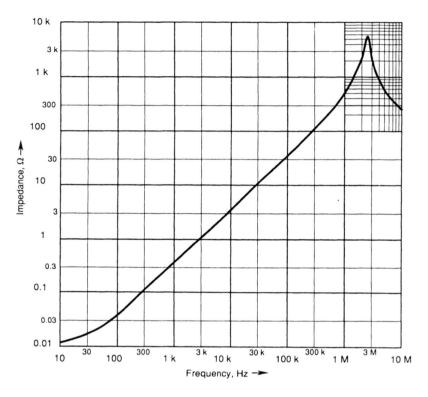

Figure 4.10 Impedance *versus* frequency for a 55 μH inductor. Note that the logarithmic scales are divided in multiples of 3 and 10.

Parallel Capacitance, C_p:

The parallel resonant frequency is 2.4 MHz. We can solve for the parallel capacitance from the relationship:

$$\omega_0 \approx \frac{1}{\sqrt{LC}} \qquad (2)$$

$$C_p = \frac{1}{L(\omega_0)^2} \qquad (3)$$

$$= \frac{1}{55.7 \times 10^{-6} \times (2 \times \pi \times 2.4 \times 10^6)^2} \tag{4}$$

$$= 79 \text{ pF}$$

C_p is the result of turn-to-turn winding capacitances, and the capacitance between the winding and the magnetic core.

Parallel Resistance, R_p:

The parallel resonance forms a peak in the impedance curve with a maximum value at $f = 2.4$ MHz. The parallel resistance, R_p, is then just the value of the impedance, Z, at a resonance which is 5.45 kΩ. This resistance represents the core and winding resistance losses (including skin effect) at the resonant frequency.

Commentary and Conclusions:

1. The test data show close to ideal inductor performance from $f = 100$ Hz, up to about 1 MHz. Above this frequency, the turn-to-turn and magnetic core capacitances begin to dominate. Thus, in the application of inductors, the inductor's resonant frequency must be beyond the operating frequencies.

Chapter 5
Ancillary Circuit Elements

5.1 INTRODUCTION

Circuit board designs can require the consideration of supporting equations and concepts such as are covered in this chapter. These include conductor and ground plane resistance, voltage and current sources, and dc power supplies.

The chapter shows how to calculate conductor resistance, needed for crosstalk analysis and dc current bus voltage drops. Order of magnitude resistances for ground planes are determined.

Voltage and current sources are defined. This chapter also shows several current source implementations because of their frequent use in electronic circuits.

Finally, power supply parameters are defined and ripple voltage is discussed.

5.2 PRINTED WIRING BOARD RESISTANCE

5.2.1 Conductor Resistance, R-1

Voltage drops caused by the current through land resistances between two points on printed wiring boards can be an important factor for determining signal crosstalk and dc voltage offsets.

The resistance of the rectangular conductor with dimensions shown in Fig. 5.1 is

$$R = \rho \frac{l}{A} \tag{1}$$

$$= \rho \frac{l}{w \times t} \, \Omega \qquad \text{(dimensions in meters,} \quad \rho = \text{resistivity in } \Omega\text{-m)} \tag{2a}$$

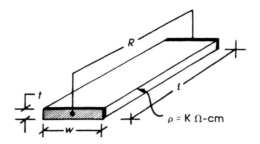

Figure 5.1 A section of a printed wiring board land.

$$= 0.394 \, \rho' \, \frac{l}{w \times t} \, \Omega \quad \begin{array}{l} \text{(dimensions in inches,} \\ \rho' = \text{resistivity in } \Omega\text{-cm)} \end{array} \quad (2b)$$

$$= \rho' \, \frac{l}{w \times t} \, \Omega \quad \begin{array}{l} \text{(dimensions in centimeters,} \\ \rho' = \text{resistivity in } \Omega\text{-cm)} \end{array} \quad (2c)$$

Example:

Determine the resistance for a copper foil PWB land which is 0.010" wide, 0.0028" thick (2 oz.) and 1.2" long. For copper at 20° C, $\rho' = 1.725 \times 10^{-6}$ Ω-cm. Using Eq. (2b):

$$R = 0.394 \times 1.725 \times 10^{-6} \times \frac{1.2''}{0.010'' \times 0.0028''}$$

$$= 0.029 \, \Omega \quad (3)$$

Commentary and Conclusions:

1. Tables generally give the resistivity, ρ' in Ω-centimeter units. This is literally the resistance between two opposing faces of a cube of the material which measures 1 cm on a side. If we used centimeters in the example, $w = 0.010''$ × 2.54 cm/in = 0.0254 cm, $t = 0.0028'' \times 2.54$ cm/in = 0.0071 cm and l = 1.2" × 2.54 cm/in = 3.05 cm. The resistance is given by Eq. (2c):

$$R = \rho' \, \frac{l}{w \times t}$$

$$= 1.725 \times 10^{-6} \times \frac{3.05 \text{ cm}}{0.0254 \text{ cm} \times 0.0071 \text{ cm}}$$

$$= 0.029 \ \Omega \tag{4}$$

which is the same result as in the example.
2. The resistivity of all materials is temperature dependent. Copper, for example, has a temperature coefficient of $+0.0039/°C$. Accordingly, the resistance for copper increases by ≈ 40 percent for a $+100°C$ temperature change.

5.2.2 Ground Plane Resistance, R-2

The numerical value of resistance between two points on a ground plane is important in low-crosstalk electronic circuits. This is because this value can be used to assess the effectiveness of the ground plane in reducing the mutual capacitance or inductance between channels or signals. Figure 5.2 shows a typical ground plane.

The resistance-unit thickness between the two conductors on the ground plane is given by

$$R \times t = \frac{\rho'}{100\pi} \ln\left(\frac{d}{r}\right) \Omega\text{-m} \tag{1a}$$

$$= 0.216 \times 10^{-6} \ln\left(\frac{d}{r}\right) \Omega\text{-in (for copper)} \tag{1b}$$

$$= 0.549 \times 10^{-6} \ln\left(\frac{d}{r}\right) \Omega\text{-cm (for copper)} \tag{1c}$$

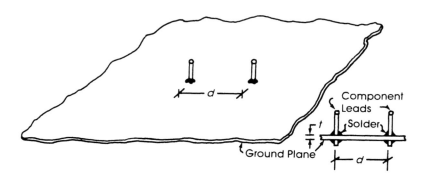

Figure 5.2 An infinite ground plane (with regard to conductor spacing) with two component leads soldered to it in plated-through holes.

where

$$\rho' = \text{resistivity}, \Omega\text{-cm}$$

The following three examples are designed to give order-of-magnitude values for ground plane resistance.

Example 1:

Calculate the resistance between two component leads soldered in 0.035" diameter plated-through holes in a 0.0028" thick ground plane 1.2" apart. Using Eq. (1b), we get

$$R = \frac{0.216 \times 10^{-6}}{0.0028''} \ln\left(\frac{1.2''}{0.035''/2}\right) \Omega$$

$$= 326 \times 10^{-6} \, \Omega \tag{2}$$

which is a very small value.

Example 2:

In practice, an infinite conductor plane, of course, is not realizable. As the ground plane becomes finite, the resistance will be larger than that indicated by Eq. (2) above. If the conductors were mounted on the edge of a ground plane where the length and the width of the ground plane were large compared to the spacing, as shown in Fig. 5.3, the resistance between the points would be exactly double that predicted by Eq. (2), which would still be a very small value.

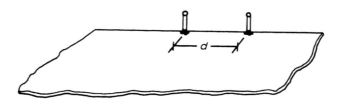

Figure 5.3 Half an infinite ground plane with conductors soldered to the edges.

Example 3:

Here the ground plane is assumed to be the width of the plated through-hole diameter as shown in Fig. 5.4.
From Subsection 5.2.1,

$$R = 0.394 \, \rho' \, \frac{l}{w \times t} \, \Omega \qquad (3)$$

$$= 0.394 \times 1.725 \times 10^{-6} \times \frac{1.2''}{0.035'' \times 0.0028''}$$

$$= 0.0083 \, \Omega \qquad (4)$$

Example Summary:

From the examples we get three values: 0.00033 Ω for the infinite plane, 0.00066 Ω for the infinite half-plane, and 0.0083 Ω for the narrow strip shown in Fig. 5.4, for an overall ratio of 25. It is recommended that the two extreme values be calculated for the geometry of interest and that these numbers be evaluated for importance to circuit operation.

Another consideration is the skin effect. Reference [1] gives the depth of penetration, δ, for copper as

$$\delta = \frac{6.6 \times 10^{-2}}{\sqrt{f}} \, \text{m} \qquad (5)$$

$$= 6.6 \times 10^{-5} \, \text{m} \qquad \text{for} \quad f = 1 \text{ MHz} \qquad (6)$$

$$= 0.0026''$$

This means that at a depth of 0.0026", the current density is $1/e$ (36.8%) that at the surface, the implication being that the resistance is increased. Thus, for $t = 0.0028''$, we would measure somewhat greater values of resistance at $f = 1$ MHz, than predicted above.

Figure 5.4 For illustrative purposes the ground plane is reduced to an absolute minimum.

Derivation:

From Section 1.4, The $LCRZ_0$ Analogy, the resistance-unit length is related to the geometrical factor, Γ, and the resistivity, ρ, in Ω-m by

$$R \times l = R \times t = \frac{\rho}{\Gamma} \tag{7}$$

and the capacitance per unit length is related to Γ by

$$\frac{C}{l} = \varepsilon \Gamma \text{ F/m} \tag{8}$$

Because the ground plane was considered to be infinite with regard to the conductor spacing and is of uniform material thickness, this configuration can be assumed to be homogeneous. Thus, we can combine Eqs. (7) and (8) to eliminate Γ:

$$R \times l = R \times t$$

$$= \frac{\rho \varepsilon}{(C/l)} \; \Omega\text{-m} \tag{9}$$

Formula Set C-1 gives C/l as

$$\frac{C}{l} \approx \frac{\pi \varepsilon_r \varepsilon_0}{\ln\left(\dfrac{d}{r}\right)} \text{ F/m} \tag{10}$$

Combining Eqs. (9) and (10), we get

$$R \times t = \frac{\rho}{\pi} \ln\left(\frac{d}{r}\right) \Omega\text{-m} \tag{11}$$

Since resistivity is generally given in Ω-cm, Eq. (11) becomes Eq. (1):

$$R \times t = \frac{\rho'}{100\pi} \ln\left(\frac{d}{r}\right) \Omega\text{-m}$$

Commentary and Conclusions:

The examples show that the ground plane resistance is very small when compared to the component impedances commonly used in electronic circuits. Thus, the ground planes can be considered, for our purposes, as equipotential surfaces. In low-level circuits, however, the impedance of lands and ground planes cannot be neglected. See Crosstalk Analysis CTG-1.

REFERENCES

1. Johnk, C.T.A., *Engineering Electromagnetic Fields and Waves*, New York, John Wiley and Sons, 1973, pp. 449, 450.

5.3 VOLTAGE SOURCES, V-1

Voltage sources are considered to be ideal voltage generators which can provide either direct, or alternating voltages, or both. By definition, a true voltage source has zero-impedance, has zero-load regulation and can provide infinite current to a short circuit. Figure 5.5 illustrates three classes of voltage sources.

dc Source. The output is a pure dc voltage.

ac Source. $A_1 \sin \omega t$ represents the fundamental frequency. The other terms can be harmonics of the fundamental, or other frequencies plus random noise.

Combined ac and dc Sources. The alternating frequencies can be power-supply ripple plus random noise superimposed on a dc voltage.

Commentary and Conclusions:

1. The concept of ideal voltage sources is useful as a starting point in circuit analysis and is frequently used to represent power supplies and operational amplifiers. Resistance, inductance, or capacitance must be added to the ideal voltage source in order to accurately model actual circuit elements.
2. Other types of voltage sources are available. These include saw-toothed, square-wave, and ramp generators, for example.

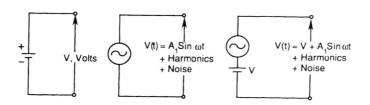

Figure 5.5 Voltage sources can deliver direct current, alternating current, or a combination of the two.

5.4 CURRENT SOURCES AND SINKS, I-1

Current sources are considered to be ideal current generators, which can provide a predetermined current regardless of the load conditions. A current sink is a type of load in which the current is independent of the applied voltage. Figure 5.6 illustrates these two concepts.

Current Source Example:

Figure 5.7 shows the circuit diagram for a simple current source constructed with a *pnp* bipolar transistor. The voltage across resistor R_1 is proportional to the emitter current of Q_1. This voltage is compared with that of Zener diode, CR_1 (less the Q_1 emitter-base voltage). Because the collector current of Q_1 almost equals its emitter current, the output current, i_1, is approximately regulated and is thus semi-independent of the load impedance value.

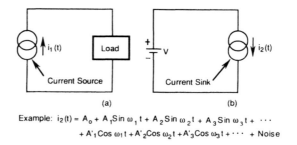

Example: $i_2(t) = A_o + A_1 \sin \omega_1 t + A_2 \sin \omega_2 t + A_3 \sin \omega_3 t + \cdots$
$+ A'_1 \cos \omega_1 t + A'_2 \cos \omega_2 t + A'_3 \cos \omega_3 t + \cdots + \text{Noise}$

Figure 5.6 The current source (a) provides a constant current to the load. The current sink (b) draws current from the supply, V. This current may be pure dc or contain alternating currents of various frequencies.

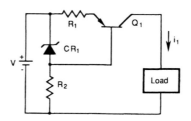

Figure 5.7 This circuit provides nearly constant current to the load within the linear operating range of transistor Q_1.

Current Sink Examples:

Current sinks can be constructed using *npn* transistors as shown in Fig. 5.8a.

Commentary and Conclusions:

1. Current sources are occasionally implemented in practice. The concept of current sinks is very useful in circuit analysis because it provides a means of quickly calculating the effect of load currents on other circuit elements. Into one current sink we can lump all the currents due to the various loads, and then we may investigate the effect of these currents on line and bus impedance voltage drops.

5.5 POWER SUPPLY CHARACTERISTICS, PS-1

Ideally, dc power supplies would, at their output terminals, be pure voltage sources with zero impedance and no ripple voltage, as shown in Fig. 5.9.

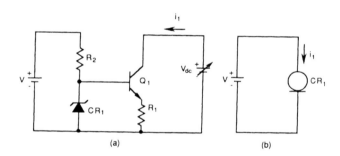

Figure 5.8 Using the same principles as above, the current i_1 (a) is approximately independent of the impressed voltage, V_{dc}. The current-regulating diode (b), shown symbolically, has the advantage of providing a current sink with a single passive component.

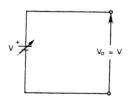

Figure 5.9 This supply provides pure dc.

Real power supplies have nonzero output impedance and contain ripple voltage. The impedance can increase crosstalk and the ripple voltage can produce in-band noise, which may either be mistaken for or mask the signal of interest.

Description:

The various components shown in Fig. 5.10 are described:

V_{dc} = voltage equivalent to an ideal battery;

R_b = equivalent output resistance of the regulator portion of the power supply. This resistance may actually have a *negative* value due to remote sensing and other feedback loop characteristics;

C = power supply output capacitor;

L_C = inductance of the output capacitor;

R_C = equivalent series resistance of the output capacitor;

L_0 = inductance of leads between the output capacitor and the power supply terminals;

R_0 = resistance of the leads between the output capacitor and power supply terminals;

V_n = output "ripple" voltage. Sometimes this quantity is specified as x mV rms which can be misleading.

Figure 5.11 shows a 15 mV rms output ripple.

Commentary and Conclusions:

1. This discussion is intended as a very short overview of regulated power supplies, as related to crosstalk and noise issues. More detailed descriptions are available from manufacturers and publications.
2. A wide range of power supply ratings and types are available and each application will have its own particular requirements.

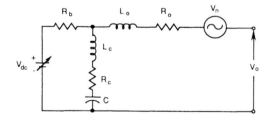

Figure 5.10 All realizable dc power supplies comprise the elements shown in this figure. The relative values depend on the type and quality of the individual power supply.

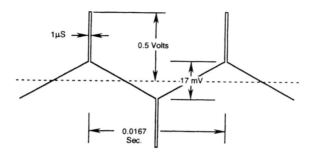

Figure 5.11 The 0.5 V spikes are riding on a 17 mV p-p triangular wave. The true rms value is 15 mV. However, the 0.5 V spikes may introduce crosstalk into sensitive circuits and produce undesirable effects.

Chapter 6
Experiments and Test Data

6.1 INTRODUCTION

Experimental data provide valuable confirmation for theoretical analyses. Comparison between "should be" predictions and the test data yields these benefits:

- Validates the ideas and equations presented in the Formula Sets and Crosstalk Analyses;
- Provides deeper insight into the physical effects involved when predictions do not match data;
- Enhances overall understanding;
- Uncovers second-order effects;
- Enhances instrumentations skills.

For these reasons, we designed a set of experiments and constructed two circuit boards for measurement purposes.

6.1.1 Experimental Results

The tests conducted were very successful with good to excellent correlation on all experiments. Differences between predicted and measured results ranged from 0 dB to 4.7 dB, accuracies more than adequate for most crosstalk needs. The experimental results are summarized in Table 6.1.

6.1.2 Circuit Board Hardware Design

Two almost identical printed wiring boards were fabricated. Photos of these boards appear in Figs. 6.1 through 6.4.

Table 6.1
Experimental Results—Predicted *versus* Measured Crosstalk

Experiment	Formula Set or Analysis	Predicted Crosstalk (dB)	Measured Crosstalk (dB)
EXP C-4	C-4	+12.9	+13.2
EXP C-5A	C-5	+1.4	+2.5
EXP C-5B.1	C-5	−1.1	−2.6
EXP C-5B.2	C-5	+16.8	+13.4
EXP C-7	C-7	−24.1	−28.0
EXP L-8 Case I	L-8	−36.2	−33.8
EXP L-8 Case II	L-8	−56.2	−56.7
EXP L-8 Case III	L-8	+11.9	+9.8
EXP CTG-1A	CTG-1	−33.5	−28.8
EXP CTG-1B	CTG-1	−42.0*	−42.4
EXP CTG-1C	CTG-1	−42.0*	−38.0

*Noise Floor

6.1.3 Circuit Board Schematics

The principal amplifier for both circuit boards A and B are identical. The auxiliary amplifiers are shown in Fig. 6.6.

6.1.4 Test Equipment

Standard laboratory supplies and oscillators were used to provide the 15 V input power and sinusoidal signals, respectively. High-level signals were measured with a Tektronix dual-channel oscilloscope. A Hewlett-Packard model HP 3825 spectrum analyzer was used to measure low level signals.

6.2 CAPACITIVE CROSSTALK

6.2.1 Capacitance between Parallel, Vertical, Flat Conductors, EXP C-4

This is a geometrical relationship that occurs frequently on printed wiring boards. This experiment showed a very close correlation between calculated and experimental results:

$$\text{measured crosstalk} = 13.2 \text{ dB}$$

$$\text{predicted crosstalk} = 12.9 \text{ dB}$$

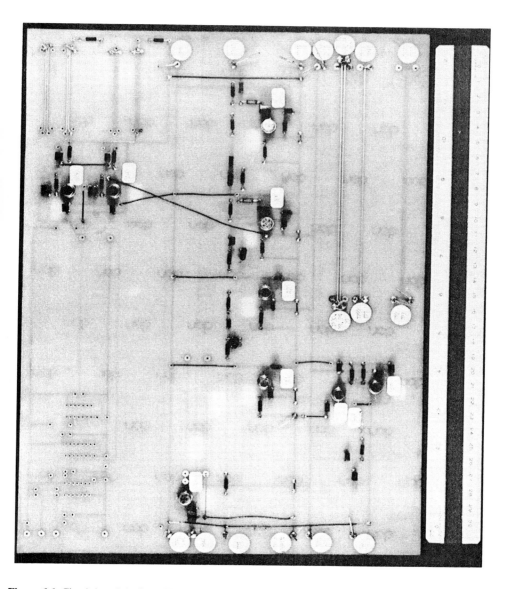

Figure 6.1 Circuit board A, front. The component side of this board shows the open construction designed specifically for crosstalk measurements. The center section of the board contains the principal amplifier, which, between test points TP4 and TP7, and TP13 and TP14, has an overall gain of 1051 (calculated at 1099). Auxiliary amplifiers appear in the lower left and upper right corners. Parallel lands, 6″ long, in the lower right corner are used for capacitive and mutual inductance measurements.

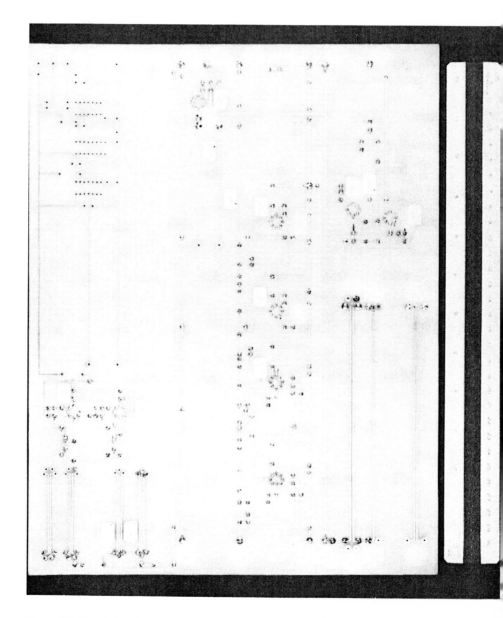

Figure 6.2 Circuit board A, back. The lands shown connect the operational amplifiers, resistors, and capacitors forming the various circuits. Located in the upper left are two sets of three lands used in EXP C-5B.1, guard rings.

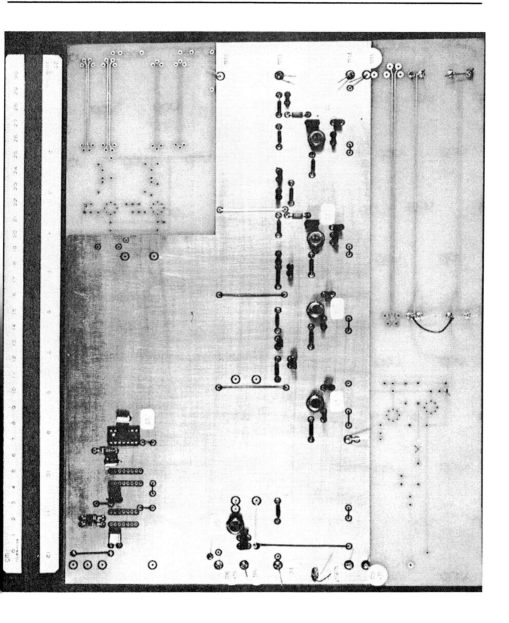

Figure 6.3 Circuit board B, front. A ground plane has been added to the component side of the circuit board. The pricipal amplifier is the same as that on circuit board A with the same measured gain of 1051. An auxiliary amplifier appears in the upper left corner. The lower right corner contains parallel 6″ lands and test pins for measuring land-to-land capacitance and inductance.

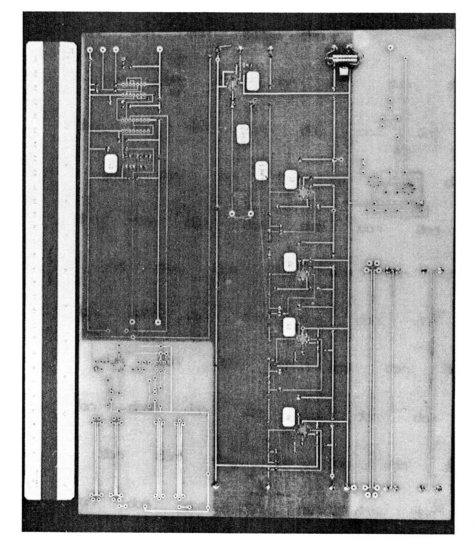

Figure 6.4 Circuit board B, back. The land pattern is identical to Board A.

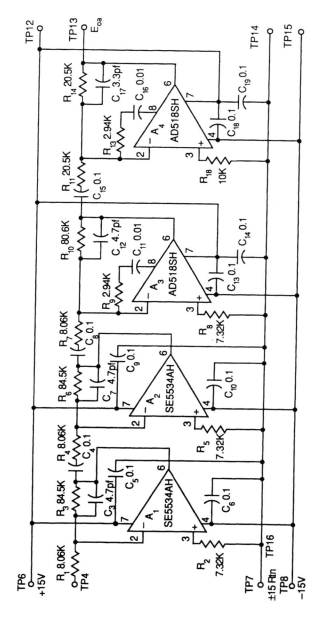

Figure 6.5 The principal amplifiers contain four stages, A_1 through A_4, with calculated gains of 10.5, 10.5, 10, and 1, respectively. Operational amplifiers are of types SH 5534AH for stages 1 and 2, and AD 518SH for stages 3 and 4.

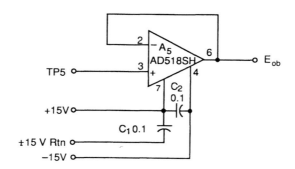

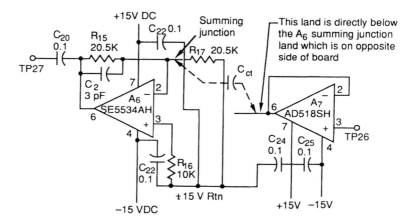

Figure 6.6 Auxiliary amplifiers used are A_5 through A_8, and are located as shown in Figs. 6.1 and 6.3. These use operational amplifier types of SE 5534AH and AD 528SH, as noted on the schematics. Amplifiers A_9 and A_{10} were not used in these experiments.

Test Circuit:

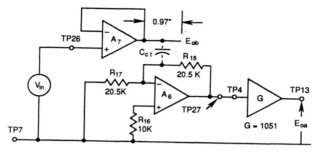

Figure 6.7 The output of amplifier A_7 has a trace, 0.010″ wide × 0.97″ long, running directly above the summing junction for amplifier A_6. The capacitance between these two traces provides an input to amplifier A_6. For measurement purposes, the output of A_6 is amplified by 1051×.

Discussion:
1. The amplifier layout is shown in Fig. 6.8.
2. Figure 6.9 shows the land-to-land edge view drawn to scale.
3. The narrowness of traces, the large separation, and the short length may intuitively suggest very little coupling. Yet, this circuit has a unity-gain crosstalk level of −48 dB.
4. This experiment clearly illustrates the fringing flux concept, and shows that the actual capacitance is 4.0 × that which would be calculated from parallel plates neglecting fringing.

Hardware Configuration:

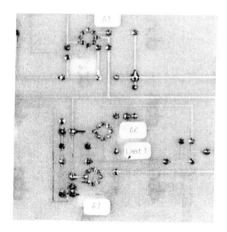
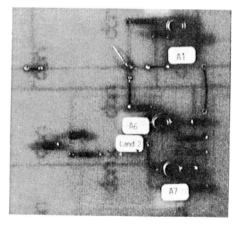

Figure 6.8 These photographs show the output of amplifier A_7 connected to land 1(a). Land 2(b) is directly above land 1. Land 2 is cut near the "×."

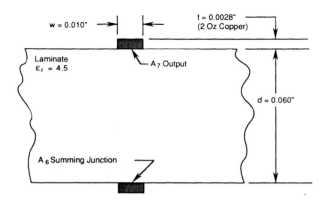

Figure 6.9 Test circuit board, edge view drawn to scale.

Measured Crosstalk:

For the circuit shown in Fig. 6.7, we obtained the following data:

$$E_{oa} = 4.55 \text{ V} \qquad f = 50 \text{ kHz}$$
$$E_{ob} = V_{in} = 1.00 \text{ V} \qquad G = 1051$$
$$R_{17} = 20.5 \text{ k}\Omega$$

From the measured data, the crosstalk is then

$$\frac{E_{oa}}{E_{ob}} = \frac{4.55 \text{ V}}{1.00 \text{ V}} = 4.55 \; (K_C = +13.2 \text{ dB}) \tag{1}$$

Predicted Crosstalk:

Crosstalk Analysis CTC-1 gives the crosstalk, K_C, in dB, as

$$K_C = 20 \log\left(\frac{E_{oa}}{E_{ob}}\right) = 20 \log(2\pi f R_1 C_{ct} G) \text{ dB} \tag{2}$$

C_{ct}, calculated from Formula Set C-4, is

$$\frac{C_{ct}}{l} = 0.225 \varepsilon_r K_{C1}\left(\frac{w}{d}\right) \text{pF/in} \tag{3}$$

$$\frac{d}{w} = \frac{0.060''}{0.010''} = 6 \tag{4}$$

The fringing factor, K_{C1}, is determined from Fig. 6.10. C_{ct} is then

$$C_{ct} = 0.225 \times 4.5 \times 4.0 \times \left(\frac{0.010''}{0.060''}\right) \times 0.97'' \text{ pF}$$

$$= 0.65 \text{ pF} \tag{5}$$

Noting that R_{17} is the equivalent of R_1 in Eq. (2), the predicted crosstalk is then

$$K_c = 20 \log\left(\frac{E_{oa}}{E_{ob}}\right) = 20 \log(2\pi \times 0.050 \times 10^6 \times \ldots$$

$$\ldots \times 20.5 \times 10^3 \times 0.65 \times 10^{-12} \times 1.051 \times 10^3) \text{ dB}$$

$$= +12.9 \text{ dB} \tag{6}$$

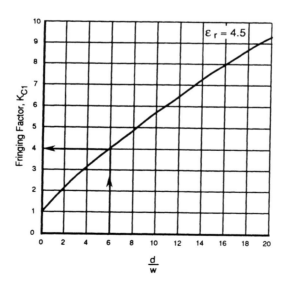

Figure 6.10 For $d/w = 6$, the fringing factor is 4.0.

6.2.2 Capacitance between Parallel, Horizontal, Flat Conductors, EXP C-5A

This experiment measures the crosstalk between two parallel lands 3" long, 0.5" apart and on the same layer of a printed circuit board. This experiment showed good correlation between measured and predicted crosstalk:

$$\text{measured crosstalk} = 2.5 \text{ dB}$$
$$\text{predicted crosstalk} = 1.4 \text{ dB}$$

Test Circuit:

The parallel traces on the test circuit board are connected as shown in Figure 6.11 below.

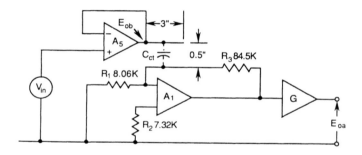

Figure 6.11 The output E_{ob}, of amplifier A_5 provides a direct input to the summing junction of amplifier A_1 through the crosstalk capacitance C_{ct} as if it were a discrete component.

Discussion:

1. As in EXP C-4, the narrowness of the traces and the relatively large separation suggest very little crosstalk, but the unity-gain level is actually -59 dB when using the circuit values and operating frequency for this experiment.
2. Figure 6.12 is a photograph of the land configuration. Figure 6.13 shows the land-to-land edge view drawn to scale.

Hardware Configuration:

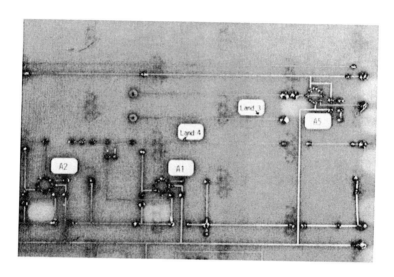

Figure 6.12 Land 3 is connected to the output of amplifier A_5 and land 4 is connected to the summing junction of A_1. The lands are 3" long, separated by 0.5".

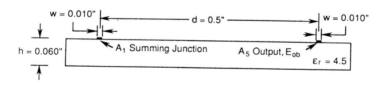

Figure 6.13 Two parallel 0.010" lands, 0.5" on centers, are on a 0.060" thick printed circuit board having a relative dielectric constant of 4.5.

Measured Crosstalk:

For the circuit shown in Fig. 6.11 the data is

$$E_{oa} = 1.34 \text{ V} \qquad f = 50 \text{ kHz}$$
$$E_{ob} = V_{in} = 1.00 \text{ V} \qquad G = 1051$$
$$R_1 = 8.06 \text{ k}\Omega$$

From the measured data, the crosstalk is then:

$$\frac{E_{oa}}{E_{ob}} = \frac{1.34 \text{ V}}{1.00 \text{ V}} = 1.34 \ (K_c = 2.5 \text{ dB}) \tag{1}$$

Predicted Crosstalk:

From Crosstalk Analysis CTC-1, the crosstalk, K_c, is given by

$$K_c = 20 \log\left(\frac{E_{oa}}{E_{ob}}\right) = 20 \log(2\pi f R_1 C_{ct} G) \text{ dB} \tag{2}$$

C_{ct}, is calculated from Formula Set C-5 as

$$\frac{C_{ct}}{l} = \frac{0.71 \ \varepsilon_{r(\text{eff})}}{\ln\left(\dfrac{\pi d}{w + t}\right)} \text{ pF/in} \tag{3}$$

For

$$l = 3'' \qquad w = 0.010''$$
$$d = 0.5'' \qquad t = 0.0028''$$
$$h = 0.060'' \qquad \varepsilon_{r(\text{eff})} \approx 1 \ (\text{because } d/h = 8.3)$$

$$C_{ct} = \frac{0.71 \times 1 \times 3''}{\ln\left(\dfrac{\pi \times 0.5''}{0.010'' + 0.0028''}\right)} \text{ pF}$$

$$= 0.44 \text{ pF} \tag{4}$$

The predicted crosstalk is then:

$$K_c = 20 \log\left(\frac{E_{oa}}{E_{ob}}\right) = 20 \log(2\pi \times 0.050 \times 10^6 \times \ldots$$

$$\ldots \times 8.06 \times 10^3 \times 0.44 \times 10^{-12} \times 1.051 \times 10^3) \text{ dB}$$

$$= +1.4 \text{ dB} \tag{5}$$

6.2.3 Horizontal Flat Conductors with Guard Rings, EXP C-5B

Guard rings are sometimes used to reduce crosstalk to low-level input circuits. The rings, connected to circuit ground, "draw" the electric flux to themselves, and thus help protect the sensitive input circuit from interfering signals. In this case, the guard rings are circuit board lands. Two experiments were conducted, first with guard lands and then with the lands connected in parallel. This experiment indicates a 16 dB crosstalk improvement by the use of guard lands. (On an equal basis the crosstalk improvement is actually 9.9 dB; see Commentary and Conclusions, item 3.)

With guard lands (Exp C-5B.1):

$$\text{measured crosstalk} = -2.6 \text{ dB}$$
$$\text{predicted crosstalk} = -1.1 \text{ dB}$$

Without guard lands (Exp C-5B.2):

$$\text{measured crosstalk} = 13.4 \text{ dB}$$
$$\text{predicted crosstalk} = 16.8 \text{ dB}$$

Test Circuit:

In this experiment, both the interfering and "victim" lands have guard lands positioned on each side as shown in Fig. 6.14.

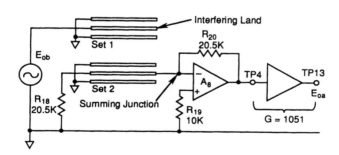

Figure 6.14 Two sets of three parallel lands are placed on the circuit board. The center conductor of land set 1 is connected to a voltage source, E_{ob}. In land set 2, the center conductor is connected to the summing junction of A_8. The other four lands are grounded.

Discussion:

1. Figure 6.15 shows an edge view of the two land sets.
2. Crosstalk measurements were made and compared with predicted results.
3. To provide additional data for capacitive crosstalk, the land sets were connected in parallel as shown in Fig. 6.16.
4. The crosstalk was remeasured, compared with the predicted value.

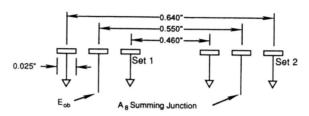

Figure 6.15 Two land sets, each with three 0.025" conductors separated by 0.020".

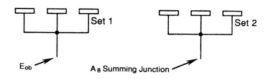

Figure 6.16 Land set 1 is connected to E_{ob}; set 2 to the A_8 summing junction.

Hardware Configuration (Figs. 6.17 and 6.18):

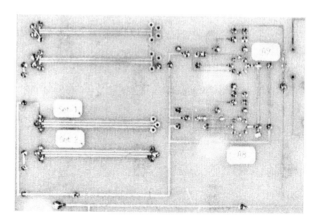

Figure 6.17 This photograph shows land sets 1 and 2; as noted, the center land of set 1 is connected to E_{ob}. The center land of set 2 is connected to amplifier A_8 summing junction.

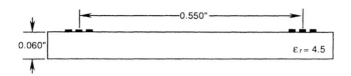

Figure 6.18 Two 2" long sets of three parallel 0.025" lands, each separated by 0.020". The lands are 0.0028" thick on a 0.060" epoxy-glass board ($\varepsilon_r = 4.5$).

Measured Crosstalk (EXP C-5B.1):

For the circuit shown in Fig. 6.14, the data are

$$E_{oa} = 0.74 \text{ V} \qquad f = 50 \text{ kHz}$$
$$E_{ob} = 1 \text{ V} \qquad G = 1051$$
$$R_{18} = 20.5 \text{ k}\Omega$$

The measured crosstalk is

$$\frac{E_{oa}}{E_{ob}} = \frac{0.74 \text{ V}}{1.00 \text{ V}} = 0.74 \ (K_c = -2.6 \text{ dB}) \tag{1}$$

Predicted Crosstalk:

As noted, the outside conductors are grounded as shown in Fig. 6.19. Several flux lines have been added to illustrate the ideas presented.
  We can make these observations:
1. The grounded "sending" conductors do not produce electric flux.
2. The low output impedance operational amplifier producing E_{ob} impresses voltage on land 2. Hence, the voltage on land 2 is independent of all other land potentials.

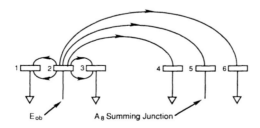

Figure 6.19 Flux generated by the E_{ob} potential on land 2 travels to lands 1, 3, 4, 5, and 6.

3. Land 2 provides electric flux to guard lands 1 and 3. These flux lines do not materially contribute to the value of C_{ct}.
4. By virtue of the potential on land 2, electric flux is "sent" to lands 4, 5, and 6.
5. Lands 4, 5, and 6 are all at essentially ground potential; lands 4 and 6 are connected to ground, and land 5 is connected to the A_8 summing junction, a virtual ground.
6. The flux leaving land 2 is thus divided into three parts, one each to lands 4, 5, and 6.
7. The land's effective area is reduced because of the flux distributions. Thus, for land 2, the area is $2w$; for land set 2, the area is $6w + 2t$. Please see Measured Crosstalk (EXP C-5B.2).

Because land 2 is 0.025" wide and lands 4, 5, and 6 have a combined width of 0.075", Eq. (6b) from Formula Set C-5 is used:

$$\frac{C}{l} \approx \frac{1.41\, \varepsilon_{r(\text{eff})}}{\ln\left[\pi^2 d^2 \left(\dfrac{1}{w_1 + t}\right)\left(\dfrac{1}{w_2 + t}\right)\right]}\, \text{pF/in} \qquad (2)$$

As noted in observation 7, the effective values are $w_1 = 0.025''$, $t_1 = 0''$, $w_2 = 0.075''$, $t_2 = 0.0028''$. The ratio d_{ave}/h is $0.0550''/0.060'' = 9.2$; thus, $\varepsilon_{r(\text{eff})} \approx 1$. For $l = 2''$, the total capacitance between land 2 and lands 4, 5, and 6 is

$$C \approx \frac{1.41 \times 1 \times 2''}{\ln\left[\pi^2 (0.550'')^2 \left(\dfrac{1}{0.025''}\right)\left(\dfrac{1}{0.075'' + 0.0028''}\right)\right]}\, \text{pF}$$

$$\approx 0.39\, \text{pF} \qquad (3)$$

Because only $1/3$ of the total flux reaches land 4, which is connected to the summing junction of A_8, the crosstalk capacitance is

$$C_{ct} = 0.13\, \text{pF} \qquad (4)$$

Noting that R_{18} is equivalent to R_1, the predicted crosstalk is

$$K_c = 20 \log\left(\frac{E_{oa}}{E_{ob}}\right) = 20 \log(2\pi \times 0.050 \times 10^6 \times 20.5 \times 10^3 \times \ldots$$
$$\ldots 0.13 \times 10^{-12} \times 1.051 \times 10^3)\, \text{dB}$$
$$= -1.1\, \text{dB} \qquad (5)$$

Measured Crosstalk (Lands Paralleled: EXP C-5B.2)

For the lands connected as in Fig. 6.16, the test data was:

$$E_{oa} = 4.7 \text{ V} \qquad f = 50 \text{ kHz}$$
$$E_{ob} = 1.0 \text{ V} \qquad G = 1051$$
$$R_{18} = 20.5 \text{ k}\Omega$$

The measured crosstalk is then:

$$\frac{E_{oa}}{E_{ob}} = \frac{4.7}{1.0} \quad (K_c = 13.4 \text{ dB}) \tag{6}$$

Predicted Crosstalk (Lands Paralleled):

The physical geometry complicates the calculation of the crosstalk capacitance. Consider Fig. 6.20, a sketch of the six conductors wherein the leaving and arriving flux lines are plotted.

This arrangement may be considered as three capacitors in parallel consisting of land pairs 1–6, 2–5, and 3–4.

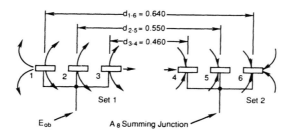

Figure 6.20 Because conductor set 1, comprising lands 1, 2, and 3, are at the same potential, no flux lines flow between them. This is also true for land set 2. Thus, flux lines leave conductor 1 on the left side, top, and bottom, and arrive at conductor 6 in a similar way (due to symmetry). For conductors 2 and 5, only the tops and bottoms have flux lines. Conductors 3 and 4 have flux lines on the tops, bottoms, and inside edges.

Formula Set C-5 gives the capacitance approximately as

$$\frac{C}{l} \approx \frac{0.71\, \varepsilon_{r(\text{eff})}}{\ln\left(\dfrac{\pi d}{w + t}\right)} \tag{7}$$

To get an approximate value for each of the three capacitors, these dimensions are used.

For land pair 1–6:
$$d = 0.640'', \quad w = 0.025'', \quad t = 0.0014''$$

For land pair 2–5:
$$d = 0.0550'', \quad w = 0.025'', \quad t = 0''$$

For land pair 3–4:
$$d = 0.460'', \quad w = 0.025'', \quad t = 0.0014''$$

The capacitance for land pair 1–6 is then

$$C_{1-6} \approx \frac{0.71 \times 1 \times 2''}{\ln\left(\dfrac{\pi \times 0.64''}{0.025'' + 0.0014''}\right)}$$

$$= 0.33 \text{ pF} \tag{8}$$

Similarly,

$$C_{2-5} \approx \frac{0.71 \times 1 \times 2''}{\ln\left(\dfrac{\pi \times 0.550''}{0.025''}\right)}$$

$$= 0.34 \text{ pF} \tag{9}$$

and

$$C_{3-4} \approx \frac{0.71 \times 1 \times 2''}{\ln\left(\dfrac{\pi \times 0.460''}{0.025'' + 0.0014''}\right)}$$

$$= 0.35 \text{ pF} \tag{10}$$

The total crosstalk capacitance is

$$C_{ct} = C_{1-6} + C_{2-5} + C_{3-4}$$
$$= 0.33 + 0.34 + 0.35$$
$$= 1.02 \text{ pF} \quad (11)$$

The predicted crosstalk is

$$K_c = 20 \log\left(\frac{E_{oa}}{E_{ob}}\right) = 20 \log(2\pi \times 0.050 \times 10^6 \times 20.5 \times 10^3 \times \ldots$$
$$\ldots 1.02 \times 10^{-12} \times 1.051 \times 10^{+3}) \text{ dB}$$
$$= +16.8 \text{ dB} \quad (12)$$

Commentary and Conclusions:

1. This experiment explains the theory behind the use of guard rings and quantifies the expected improvements. The experiment further shows that, in this case, guard rings on the transmitting lands provide little or no benefit.
2. Material from Sections 1.1, 1.2, and 1.3 have aided in understanding the physical effects behind the test results. The principles described in these sections can thus be used to analyze circuits and other configurations beyond those shown in this book.
3. The improvement is actually 9.9 dB. For a direct comparison, we would have used single 0.025" traces rather than three. This produces a crosstalk capacitance of

$$C_{ct} \approx \frac{0.71 \times 1 \times 2''}{\ln\left(\frac{\pi \times 0.550''}{0.025'' + 0.0028''}\right)}$$
$$= 0.34 \text{ pF} \quad (13)$$

For this case, the measured crosstalk voltage ratio was

$$\frac{E_{oa}}{E_{ob}} = \frac{0.74 \text{ V}}{1 \text{ V}}$$

Calculating C_{ct} from the measured values:

$$C_{ct} = \frac{10^{12}}{2\pi f R_1 G} \times \frac{E_{oa}}{E_{ob}} \text{ pF}$$

$$= \frac{10^{12}}{2\pi \times 5 \times 10^4 \times 20.5 \times 10^3 \times 1.051 \times 10^3} \times \frac{0.74 \text{ V}}{1 \text{ V}} \text{ pF}$$

$$= 0.109 \text{ pF} \tag{14}$$

resulting in a ratio of $0.34/0.109 = 3.1$ or 9.9 dB.

6.2.4 Mutual Capacitance of Two Parallel, Horizontal, Flat Conductors Near a Ground Plane, EXP C-7

This experiment demonstrates the large crosstalk reduction resulting from the use of a ground plane:

$$\text{measured crosstalk} = -28.0 \text{ dB}$$
$$\text{predicted crosstalk} = -24.1 \text{ dB}$$

Test Circuit:

The parallel traces on the test circuit board B are connected as shown in Fig. 6.21.

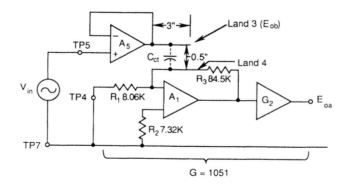

Figure 6.21 As in EXP C-5, the output of amplifier A_5 provides a direct input to the summing junction of A_1 through the crosstalk capacitance C_{ct}, as if C_{ct}, formed by lands 3 and 4, were a discrete component.

Discussion:

The hardware configuration is shown in Fig. 6.22. Figure 6.23 shows an edge view of the two lands and the ground plane drawn to scale. The presence of the ground plane reduces the calculated mutual capacitance by a factor of $0.44/0.024 = 18.7$, representing a reduction in crosstalk of 25.4 dB. As noted in Formula Set C-7, this configuration is very distance-sensitive with the mutual capacitance inversely proportional to the separation distance squared.

Hardware Configuration:

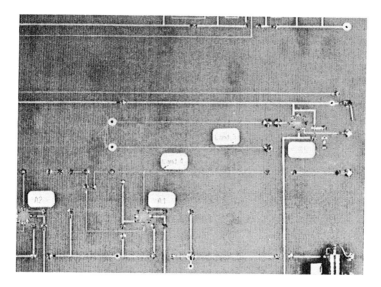

Figure 6.22 This photograph shows land 3, connected to the output of A_5, and land 4, connected to the summing junction of A_1.

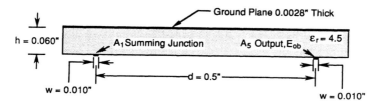

Figure 6.23 Two parallel 0.010″ lands, 0.5″ on centers, and 3″ long are on the bottom of a 0.060″ thick printed circuit board. A ground plane, with holes for component leads, covers the top (component) side of the board.

Measured Crosstalk:

For the circuit shown in Fig. 6.21, the test data are

$$E_{oa} = 0.04 \text{ Vp-p} \qquad f = 50 \text{ kHz}$$
$$E_{oa} = 1 \text{ Vp-p} \qquad G = 1051$$
$$R_1 = 8.06 \text{ k}\Omega$$

From the measured data, the crosstalk is

$$\frac{E_{oa}}{E_{ob}} = \frac{0.04 \text{ Vp-p}}{1.0 \text{ Vp-p}} = 0.04 \quad (K_c = -28.0 \text{ dB}) \tag{1}$$

Predicted Crosstalk:

From crosstalk analysis CTC-1,

$$K_c = 20 \log \left| \frac{E_{oa}}{E_{ob}} \right| = 20 \log(2\pi f R_1 C_{ct} G) \text{ dB} \tag{2}$$

In this case, $C_{ct} = C_{mct}$, the mutual crosstalk capacitance. From Formula Set C-7, we can predict the value of the mutual capacitance, C_{mct}:

$$\frac{C_{mct}}{l} = 0.07 \, \varepsilon_r K_{L1} K_{C1} \left(\frac{w}{d} \right)^2 \text{ pF/in} \tag{3}$$

From Figs. 6.21 and 6.23,

$$l = 3'', \quad d = 0.5'', \quad w = 0.010'', \quad h = 0.060''$$

$$\frac{2h}{w} = \frac{2 \times 0.060''}{0.010''} = 12, \qquad \varepsilon_r = 4.5$$

The fringing factors are determined from Fig. 6.24.

The crosstalk capacitance, C_{mct}, predicted value is

$$C_{ct} = 0.07 \times 4.5 \times 9.65 \times 6.45 \left(\frac{0.010''}{0.5''} \right)^2 \times 3'' \text{ pF}$$

$$= 0.024 \text{ pF} \tag{4}$$

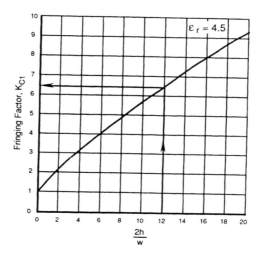

Figure 6.24 For $2h/w = 12$, the fringing factors are $K_{L1} = 9.65$ and $K_{C1} = 6.45$.

The predicted crosstalk is

$$K_c = 20 \log(2\pi \times 0.050 \times 10^6 \times 8.06 \times 10^3 \times \ldots$$
$$\ldots \times 0.024 \times 10^{-12} \times 1.051 \times 10^3) \text{ dB}$$
$$= -24.1 \text{ dB} \qquad (5)$$

6.3 INDUCTIVE CROSSTALK

6.3.1 Four-Conductor-System Mutual Inductance, L-8

In this experiment, we measured the crosstalk due to the mutual inductance between two land pairs on a printed circuit board and compared the test results with predicted values. The three cases examined, following geometries shown in Formula Set L-8 examples, are summarized in Table 6.2.

Test Circuit:

The test circuit used in each of the three cases is shown in Fig. 6.25.

Measured Crosstalk (Case 1):

In Example 2, Formula Set L-8, we calculated the mutual inductance to be 0.470 nH for the conductor arrangement shown in Fig. 6.26.

Table 6.2
Predicted *versus* Measured Crosstalk, dB

Case	L-8 Example	Predicted Crosstalk, K_l (dB)	Measured Crosstalk, K_l (dB)
I	2	−36.2	−33.8
II	3	−56.2	−56.7
III	6	11.9	9.8

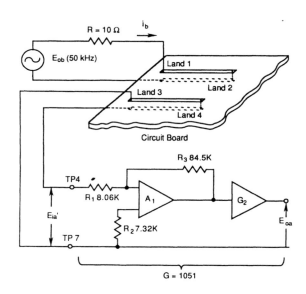

Figure 6.25 An oscillator provides a 50 kHz current to land pair 1 and 2 through a 10 Ω resistor, R. A second set of lands, 3 and 4, receive, via the mutual inductance L_m, an induced signal, E'_{ia}. This signal is first amplified by 1051× and then measured as E_{oa}. The land pairs are isolated by ground from the other circuits. Three different land-pair geometries were used as shown in Figs. 6.26, 6.27, and 6.28.

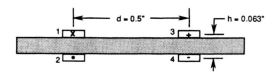

Figure 6.26 The transmitting lands are located on the left and the receiving lands on the right.

The test data for the circuit shown in Fig. 6.25 are

$$E_{oa} = 5.3 \text{ mV}_{(rms)} \qquad f = 50 \text{ kHz}$$
$$E_{ob} = 0.26 \text{ V}_{(rms)} \qquad G = 1051$$
$$R = 10 \text{ }\Omega$$

The crosstalk is then

$$\frac{E_{oa}}{E_{ob}} = \frac{0.0053 \text{ V}_{(rms)}}{0.26 \text{ V}_{(rms)}} = 0.02 \quad (K_l = -33.8 \text{ dB}) \tag{1}$$

Predicted Crosstalk (Case I):

Following the approach taken in Crosstalk Analysis CTL-1, these equations can be written from the circuit diagram:

$$E_{oa} = GE'_{in} \tag{2}$$

$$E'_{in} = L_m \frac{di_b}{dt} \tag{3}$$

$$\frac{di_b}{dt} = \frac{E_{ob}}{R} 2\pi f \tag{4}$$

Combining Eqs. (2), (3), and (4),

$$\frac{E_{oa}}{E_{ob}} = \frac{2\pi f L_m G}{R} \tag{5}$$

The crosstalk, K_l, is then:

$$K_l = 20 \log\left(\frac{E_{oa}}{E_{ob}}\right) = 20 \log\left(\frac{2\pi f L_m G}{R}\right) \text{dB} \tag{6}$$

For $L_m = 0.470$ nH, the predicted crosstalk is

$$K_l = 20 \log\left(\frac{2\pi \times 0.050 \times 10^6 \times 0.470 \times 10^{-9} \times 1.051 \times 10^3}{10}\right) \text{dB}$$
$$= -36.2 \text{ dB} \tag{7}$$

Measured Crosstalk (Case II):

In Example 3, Formula Set L-8, the mutual inductance L_m was calculated to be 0.047 nH for the geometry shown in Fig. 6.27.

In this example the crosstalk was measured as

$$\frac{E_{oa}}{E_{ob}} = \frac{0.380 \text{ mV}_{(rms)}}{260 \text{ mV}_{(rms)}} = 0.00146 \ (K_l = -56.7 \text{ dB}) \tag{8}$$

Predicted Crosstalk (Case II):

From Formula Set L-8, Example 3, L_m is 0.047 nH. The predicted crosstalk is then

$$K_l = 20 \log\left(\frac{2\pi \times 0.050 \times 10^6 \times 0.047 \times 10^{-9} \times 1.051 \times 10^3}{10}\right) \text{dB}$$
$$= -56.2 \text{ dB} \tag{9}$$

Measured Crosstalk (Case III):

Example 6 examined the case where the receiving lands 3 and 4 were enclosed by the transmitting lands 1 and 2. As would be expected, this configuration produces the highest L_m value of those considered. In the experiment we used the spacing shown in Fig. 6.28. The crosstalk was measured as

$$\frac{E_{oa}}{E_{ob}} = \frac{0.31 \text{ Vp-p}}{0.10 \text{ Vp-p}} = 3.1 \quad (K_l = 9.8 \text{ dB}) \tag{10}$$

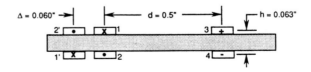

Figure 6.27 The high-level current is divided into two paths, lands 1 and 1', with the return current flowing in lands 2 and 2'.

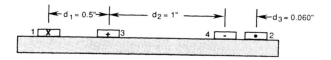

Figure 6.28 The transmitting lands enclose the receiving lands.

Predicted Crosstalk (Case III):

From Eq. (2b), Formula Set L-8, we can determine the value of L_m for the geometry shown in Fig. 6.28:

$$\frac{L_m}{l} = 0.005 \ln\left(\frac{D_{14} \times D_{23}}{D_{13} \times D_{24}}\right) \mu H/in \tag{11}$$

For this case,

$$D_{14} = 1.5'', D_{23} = 1.06'', D_{13} = 0.5'', D_{24} = 0.06'', l = 6''$$

$$L_m = 0.005 \ln\left(\frac{1.5'' \times 1.06''}{0.5'' \times 0.06''}\right) \times 6'' \, \mu H$$

$$= 119 \text{ nH} \tag{12}$$

The predicted crosstalk is then

$$K_l = 20 \log\left(\frac{2\pi \times 0.050 \times 10^6 \times 119 \times 10^{-9} \times 1.051 \times 10^3}{10}\right) dB$$

$$= +11.9 \text{ dB} \tag{13}$$

Commentary and Conclusions:

1. Table 6.2 shows very good correlation between predicted and measured results with accuracy levels that are more than adequate for most crosstalk applications.
2. Because of the low signal levels, the measurements were difficult to make, and required the use of a spectrum analyzer. Furthermore, instrumentation lead placement was critical because of crosstalk between the oscillator circuit and the $1051\times$ gain amplifier.

6.4 GROUND RETURN CROSSTALK

6.4.1 Shared-Ground Crosstalk, EXP CTG-1A

This experiment demonstrates that crosstalk can occur between signal channels due to common ground coupling. This is caused by attaching operational amplifier inputs at various locations along the ground return bus. Crosstalk Analysis CTG-1 provides the theoretical data for this experiment. (EXP CTG-1B shows the beneficial effects of using a single-point ground.) The results of this experiment are

$$\text{measured crosstalk} = -28.8 \text{ dB}$$

$$\text{predicted crosstalk} = -33.5 \text{ dB}$$

Test Circuit (Figs. 6.29 and 6.30):

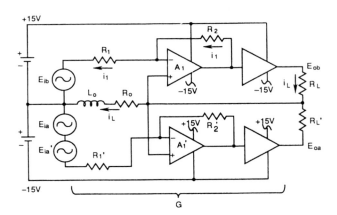

Figure 6.29 Two identical channels, A and B, are connected to common ±15 V power supplies and common ground return line. The input signal to channel B is E_{ib} and the output E_{ob}. The signal input to channel A, E_{ia}, is zero. However, a signal appears at the output of channel A due to current i_L, producing a voltage drop on ground elements L_0 and R_0 that are common to both circuits. (Current i_1 does not flow in the ground return because of the differential nature of the operational amplifier outputs.)

Hardware Configuration:

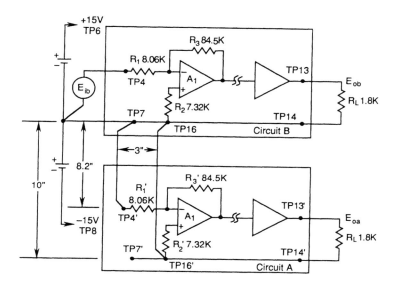

Figure 6.30 Both test circuit board assemblies, described in Section 6.1, were used in this experiment. Wires were connected between TP$_7$ and TP$_4'$, and between TP$_{16}$ and TP$_{16}'$ as shown. An input signal was applied to channel B, and the outputs of both A and B were measured.

Measured Crosstalk:

$$E_{oa} = 0.75 \text{ Vp-p} \qquad f = 50 \text{ kHz}$$
$$E_{ob} = 20.6 \text{ Vp-p} \qquad G = 1051$$
$$R_1 = 8.06 \text{ k}\Omega$$

The measured crosstalk is then:

$$\frac{E_{oa}}{E_{ob}} = \frac{0.75 \text{ V p-p}}{20.6 \text{ V p-p}} = 0.036 \ (K_g = -28.8 \text{ dB}) \qquad (1)$$

Predicted Crosstalk:

This circuit was used as an example in Crosstalk Analysis CTG-1. This analysis showed that the expected crosstalk would be

$$K_g = -33.5 \text{ dB}$$

6.4.2 Single-Point Ground Crosstalk, EXP CTG-1B

In this experiment, a single-point ground is used to reduce the crosstalk between signal channels caused by common ground coupling. This reduction is a direct result of diverting the load current from the common return bus impedance. As before Crosstalk Analysis CTG-1 provides the theoretical data for this experiment. The results of the experiment are

single-point ground measured crosstalk = -42.4 dB

common ground measured crosstalk = -28.8 dB

improvement = $+13.6$ dB

(noise floor ≈ -42 dB limits possible improvement.)

Test Circuit:

Same circuit is used as in EXP CTG-1 except ground point connections.

Hardware Configuration:

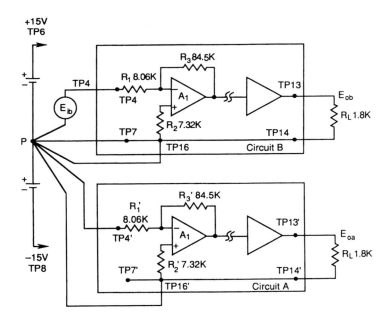

Figure 6.31 All grounds (returns) are connected at a single point, P.

Measured Crosstalk:

For the circuit shown in Fig. 6.31, we obtained these data:

$$E_{oa} = 0.16 \text{ Vp-p} \qquad f = 50 \text{ kHz}$$
$$E_{ob} = 21 \text{ Vp-p} \qquad G = 1051$$
$$R_1 = 8.06 \text{ k}\Omega$$

The measured crosstalk is then:

$$\frac{E_{oa}}{E_{ob}} = \frac{0.16 \text{ Vp-p}}{21 \text{ Vp-p}} = 0.0076 (K_g = -42.4 \text{ dB}) \tag{1}$$

Predicted Crosstalk:

As noted in Crosstalk Analysis CTG-1, theoretically, common ground crosstalk should not occur between channels if a single-point ground were used. However, in a single channel test, we measured the noise floor to be -41.9 dB. Thus, the single-point ground improved the crosstalk down to the noise floor level.

6.4.3 Downstream Power Supply Crosstalk, EXP CTG-1C

This experiment demonstrates that crosstalk between signal channels due to common-ground coupling can be reduced by locating the power supplies downstream, that is, farthest away from the low-level input stages. This relatively simple change removes the load current from the low-level input amplifier's ground bus. Crosstalk Analysis CTG-1 provides the theoretical data for this experiment. The results of this experiment are

downstream power supply measured crosstalk $= -38$ dB
common ground measured crosstalk $= -28.8$ dB
improvement $= +9.2$ dB
(noise floor ≈ -42 dB limits possible improvement.)

Test Circuit:

Same circuit is used as in EXP CTG-1 except power supply connections.

Hardware Configuration:

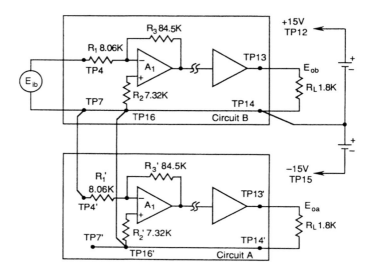

Figure 6.32 The power supply is connected at the load end of the circuit.

Measured Crosstalk:

For the circuit shown in Fig. 6.32, the measured values are

$$E_{oa} = 0.26 \text{ Vp-p} \quad f = 50 \text{ kHz}$$
$$E_{ob} = 20.7 \text{ Vp-p} \quad G = 1051$$
$$R_1 = 8.06 \text{ k}\Omega$$

The measured crosstalk is then

$$\frac{E_{oa}}{E_{ob}} = \frac{0.26 \text{ Vp-p}}{20.6 \text{ Vp-p}} = 0.013 \quad (K'_g = -38 \text{ dB}) \tag{1}$$

Predicted Crosstalk:

This circuit was also used as an example in Crosstalk Analysis CTG-1. This analysis showed that the expected crosstalk would be

$$K'_g = -106.9 \text{ dB} \qquad (2)$$

As we found in Experiment EXP GTC-1B, the -42 dB noise floor limits the improvement to -38 dB.

Appendix A
Equipotential Surface and Flux Boundary Construction for Circular Conductors

In Crosstalk Analysis, CTG-1, we saw that the knowledge of magnetic flux distribution surrounding current-carrying conductors was required to determine the induced crosstalk voltages in victim circuits. Section 1.3, Electric Field Mapping, and Formula Set C-3 show electrostatic and electromagnetic field plots for circular conductors. For both types of fields, the flux boundaries and equipotential surfaces are cylindrical. The plots were prepared by using equations provided in this appendix.

Reference [1] gives the x and y coordinates and the radii for the circles representing the cross-sectional view of the cylinders. Figure A1.1 shows the plot for an electrostatic field.

For magnetostatic fields, as shown in Fig. A1.2, the equipotential and flux tube boundary circles are interchanged.

Reference [1] develops the equations for x-axis locations and for circle radii produced by the lines of charge, equal in magnitude, opposite in polarity, and spaced a distance s apart. These are

$$\frac{d_c}{2} = \frac{s}{2}\left(\frac{A^2 + 1}{A^2 - 1}\right) \tag{1}$$

$$r_c = \frac{sA}{A^2 - 1} \tag{2}$$

$$A = \exp(2\pi\varepsilon_0\varepsilon_r\Phi/\Lambda) \tag{3}$$

where

$d_c/2$ = location on x-axis of the centers of equipotential circles with radii r_c
r_c = radius of equipotential circles

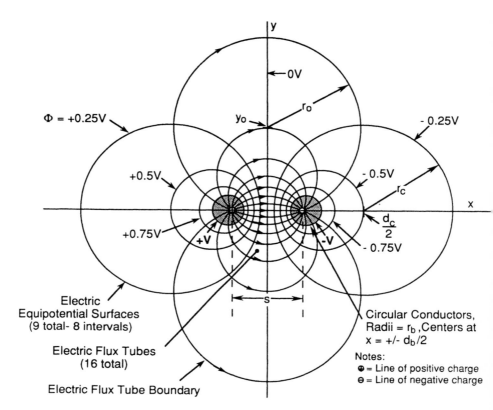

Figure A1.1 For electrostatic fields surrounding circular conductors, the equipotential surfaces have radii, r_c, with centers located on the x-axis at $x = \pm(d_c/2)$. The electric flux tube boundaries have radii, r_0, with centers on the y-axis at $\pm y_0$. Lines of charge, responsible for the generation of the electric field, are located at $x = \pm(s/2)$. Circular conductors can replace the equipotential circles, as done in Fig. 1.15.

s = center-to-center line charge distance
ε_0 = dielectric constant of vacuum, F/m
ε_r = relative dielectric constant (Ref. [1] used K)
Φ = potential of equipotential surface
Λ = linear charge density, coulombs/m

For two conductors with radii $r_c = R$, spaced distance $d_c = D$ on centers, and letting $A = A_b$, Ref. [1] (p. 90) uses Eqs. (1) and (2) to obtain expressions for s and A_b:

$$s = \sqrt{D^2 - (2R)^2} \qquad (4)$$

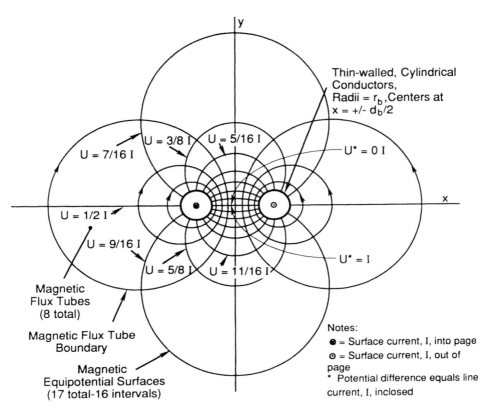

Figure A1.2 The magnetic flux tube boundaries have centers located on the x-axis, and the magnetic equipotential surfaces are centered on the y-axis. This flux plot results from currents flowing on the surfaces of thin-walled cylindrical conductors. In this example, the cylinders have radii of 2 units and are spaced 10 units on centers.

$$A_b = \frac{D}{2R} + \sqrt{\left(\frac{D}{2R}\right)^2 - 1} \tag{5}$$

Letting the potential on the conductors be $\pm\Phi_b$, Eq. (3) becomes

$$A_b = \exp(2\pi\varepsilon_0\varepsilon_r\Phi_b/\Lambda) \tag{6}$$

Taking the natural logarithm Eq. (6) and combining with Eq. (5):

$$\frac{2\pi\varepsilon_0\varepsilon_r\Phi_b}{\Lambda} = \ln(A_b)$$

$$= \ln\left[\frac{D}{2R} + \sqrt{\left(\frac{D}{2R}\right)^2 - 1}\right] \quad (7)$$

Defining

$$K_1 = \frac{2\pi\varepsilon_0\varepsilon_r}{\Lambda} \quad (8)$$

Then,

$$\Phi_b = \frac{1}{K_1}\ln(A_b) \quad (9)$$

For circles at potential Φ_c with radius r_c and spaced d_c on centers, we can get from Eqs. (3) and (8):

$$A_c = \exp(K_1\Phi_c) \quad (10)$$

or

$$\Phi_c = \frac{1}{K_1}\ln(A_c) \quad (11)$$

Dividing Eq. (11) by (9) gives the potential ratio, β:

$$\beta = \frac{\Phi_c}{\Phi_b} = \frac{\ln(A_c)}{\ln(A_b)} \quad (12)$$

Solving for A_c:

$$\ln(A_c) = \beta \ln(A_b) \quad (13)$$

or

$$A_c = (A_b)^\beta \quad (14)$$

Reference [1] (p. 74) gives the y-axis locations and radii for the electrostatic field flux lines (called *boundaries* in this appendix) as

$$y_0 = \frac{s}{2\tan k\frac{2\pi}{n}} \quad (15)$$

$$r_0 = \frac{s}{2\sin k\frac{2\pi}{n}} \quad (16)$$

where

- s = center-to-center line charge distance
- y_0 = location of the centers of the electric flux boundaries on the y-axis with radii r_0
- r_0 = radii of the flux boundary circles
- n = number of flux tubes
- k = successive integers, 1 to n

Close examination of Fig. A1.1, seen here in Fig. A1.3, shows that the number of flux tubes and equipotential surfaces have been selected so that curvilinear squares result, as discussed in Section 1.3, Electric Field Mapping. Reference [1] presents this figure as an aid to the development of curvilinear squares.

Reference [1] then presents the following equations for establishing the conditions necessary to get curvilinear squares. To be consistent with the notation used in this book, K and ε have been changed to ε_r and E, respectively.

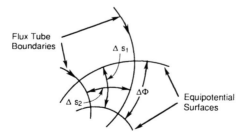

Figure A1.3 A portion of Fig. A1.1 is shown here. There are two equipotential lines representing a potential difference, $\Delta\Phi$ and two flux tube boundaries. These are spaced Δs_1 and Δs_2 apart, respectively. (Reprinted by permission of Harper & Row, Publishers, Inc.)

The electric flux density, D, is related to the electric field intensity, E, by

$$D = \varepsilon_r \varepsilon_0 E \tag{17}$$

The mean value of E is

$$E_{mean} = \frac{\Delta \Phi}{\Delta s_1} \tag{18}$$

The flux contained in each flux tube is

$$\Delta \Psi = \frac{\Lambda}{n} \tag{19}$$

The mean electric flux density is

$$D_{mean} = \frac{\Delta \Psi}{\Delta s_2} \tag{20}$$

Reference [1] combines Eqs. (17) through (20) to get

$$\frac{\Lambda}{n \Delta s_2} = \varepsilon_r \varepsilon_0 \frac{\Delta \Phi}{\Delta s_1} \tag{21}$$

and observes that, to get curvilinear squares, Δs_1 must equal Δs_2. Thus,

$$n \Delta \Phi = \frac{\Lambda}{\varepsilon_r \varepsilon_0} \tag{22}$$

Using Eqs. (1) through (7) and (15) through (22) from Ref. [1] as a foundation, we will now establish the methods for constructing the flux plot shown in Fig. A1.1.

Section 1.3 defined n_s and n_p as the number of unit capacitors in series and parallel, respectively, assuming curvilinear squares. Because n_s is equivalent to the number of equipotential intervals and n_p to the number of flux tubes, we will use this notation in the following discussion.

Assigning the potentials $\pm \Phi_b$ to the conductors shown in Fig. A1.1, observing that $n_s \Delta \Phi = 2 \Phi_b$, and letting $n = n_p$, in Eq. (22), we get

$$\frac{2 n_p \Phi_b}{n_s} = \frac{\Lambda}{\varepsilon_r \varepsilon_0} \tag{23}$$

or

$$\frac{\varepsilon_r \varepsilon_0}{\Lambda} = \frac{n_s}{2n_p \Phi_b} \quad (24)$$

Combining Eq. (24) with (7):

$$\frac{2\pi n_s \Phi_b}{2n_p \Phi_b} = \ln\left[\frac{D}{2R} + \sqrt{\left(\frac{D}{2R}\right)^2 - 1}\right] \quad (25)$$

Cancelling the common terms and rearranging:

$$\frac{n_s}{n_p} = \frac{1}{\pi} \ln\left[\frac{D}{2R} + \sqrt{\left(\frac{D}{2R}\right)^2 - 1}\right] \quad (26)$$

Equation (26) shows that the ratio n_s/n_p is independent of the conductor potential, Φ_b, the dielectric constant, $\varepsilon_r \varepsilon_0$, and the line charge, Λ, as it should be.

In Figs. A1.1 and A1.2, we deliberately selected $R = 2$ and $D = 10$ so that we would get an integral number of flux tubes and equipotential intervals. In this case,

$$\frac{n_s}{n_p} = \frac{1}{\pi} \ln\left[\frac{10}{2 \times 2} + \sqrt{\left(\frac{10}{2 \times 2}\right)^2 - 1}\right]$$

$$= 0.487 \ldots \quad (27)$$

In Fig. A1.1, we selected $n_s = 8$ and $n_p = 16$, producing the ratio $n_s/n_p = 0.5$. (To get an integral number of potential intervals and flux tubes, we can use Eq. (26) to solve for R given D. In this case, for $D = 10$ units, R must be 1.9928 ... to get the ratio 0.5.)

We can now solve for the equipotential surface radii and x-axis locations by using Eqs. (1) and (2):

$$\frac{d_c}{2} = \frac{s}{2}\left(\frac{A^2 + 1}{A^2 - 1}\right)$$

$$r_c = \frac{sA}{A^2 - 1}$$

For $R = 2$ and $D = 10$, Eq. (4) gives s as

$$s = \sqrt{D^2 - (2R)^2}$$
$$= \sqrt{(10)^2 - (2 \times 2)^2}$$
$$= 9.165 \qquad (28)$$

and A_b is from Eq. (5):

$$A_b = \frac{D}{2R} + \sqrt{\left(\frac{D}{2R}\right)^2 - 1}$$
$$= \frac{10}{2 \times 2} + \sqrt{\left(\frac{10}{2 \times 2}\right)^2 - 1}$$
$$= 4.79 \qquad (29)$$

Equation (14) gives A_c in terms of A_b and β:

$$A_c = (A_b)^\beta$$

Because we have four equipotential intervals per side, we will select $\beta = 0, 0.25, 0.5, 0.75,$ and 1. For $\beta = 1$, $A_c = A_b$ and Eqs. (1) and (2) show the $d_c/2 = 5$ and $r_c = 2$, which as they should be because the potentials of the conductor surfaces have the reference values of ± 1 V. For $\beta = 0$, we have a zero potential surface on the y-axis because $d_c/2 = \infty$ and $r_c = \infty$. The values for r_c and $d_c/2$ for values of β appear in Table A1.1.

To obtain the flux tube boundary circles and y-axis locations, Eqs. (15) and (16) are modified by substituting $n = n_p$ and letting k take the values 1, 2, 3, ..., 8.

Table A1.1
Equipotential Circle Radii and x-Axis Locations for Fig. A.1.1

β	r_c	$d_c/2$
0	∞	∞
0.25	11.4	12.3
0.5	5.3	7.0
0.75	3.1	5.5
1.0	2	5

(Because we acquire two flux tube boundaries per k value, we have duplicate values for $k = 9$ through 16, as Ref. [1] indicates.) For $k = 0$, 8, and 16, both y_0 and r_0 are infinite, representing a flux tube boundary on the x-axis. Table A1.2 lists the values.

Applying the foregoing to the problem posed in Crosstalk Analysis, CTG-1, consider the conductors shown in Fig. A1.4. Reference [1] (pp. 205–210) compares magnetic and electrostatic field plots, and shows that they are the same if the equipotential surfaces and the flux tube boundaries are interchanged. Thus, the flux ratio, α, is equivalent to the potential ratio, β. (As noted previously in Formula Set C-3 and elsewhere, the currents must flow on the conductor surfaces.)

Table A1.2
Flux Tube Boundary Circle Radii
and y-Axis Locations for
Fig. A1.1

k	r_0	y_0
0	∞	∞
1	12	11.1
2	6.5	4.6
3	5.0	1.9
4	4.6	0.0
5	5.0	−1.9
6	6.5	−4.6
7	12	−11.1
8	∞	∞

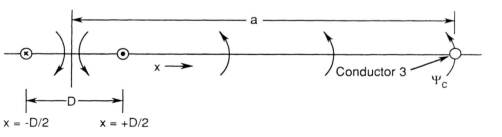

Figure A1.4 Two conductors spaced D on centers carry equal and opposite currents. A conductor (3) is located at $x = a$ and lies on the flux tube boundary Ψ_c. The fraction of the total magnetic flux linking conductor 3 produced by the conductor located at $x = +D/2$ is equal to $1 - \alpha$.

Letting $a = r_c + d_c/2$, we can show by using Eqs. (1) and (2) that

$$A_c = \frac{2a + s}{2a - s} \tag{30}$$

$$\approx \frac{2a + D}{2a - D} \quad \text{for} \quad D \gg 2R \tag{31}$$

Replacing β with α in Eq. (12),

$$\alpha = \frac{\ln(A_c)}{\ln(A_b)} \tag{32}$$

Combining Eqs. (7), (31), and (32), we get the flux ratio at $x = a$ in terms the dimensions and spacing of the conductors:

$$\alpha \approx \frac{\ln\left(\dfrac{2a + D}{2a - D}\right)}{\ln\left[\dfrac{D}{2R} + \sqrt{\left(\dfrac{D}{2R}\right)^2 - 1}\right]} \tag{33}$$

$$\approx \frac{\ln\left(\dfrac{2a + D}{2a - D}\right)}{\ln\left(\dfrac{D}{R}\right)}, \quad \text{for} \quad D \gg R \tag{34}$$

Following the methods used in Formula Set C-5, for rectangular conductors, α is given as

$$\alpha \approx \frac{\ln\left(\dfrac{2a + d}{2a - d}\right)}{\ln\left(\dfrac{\pi d}{w + t}\right)} \tag{35}$$

Thus, from the conductor dimensions their spacings, we can determine the amount of magnetic flux linking the victim circuit conductor.

REFERENCE

1. Boast, Willam B., *Principles of Electric and Magnetic Fields*, Harper and Brothers, New York, 1956, pp. 72, 74–75, 90, 205–210, 229, 311.

Bibliography

Belden Wire and Cable, *Master Catalog*, Richmond, IN, 1989.
Boast, W.B., *Principles of Electric and Magnetic Fields*, New York, Harper and Brothers, 1956.
Catt, I., "Crosstalk (Noise) in Digital Systems," *Trans. IEEE*, Vol. EC-16, No. 6, December 1967.
Catt, I., *Digital Hardware Design*, London, Macmillian, 1979.
Hoole, S., and Ratnajeevan, H., *Computer-Aided Analysis and Design of Electromagnetic Devices*, New York, Elsevier, 1989.
Howe, H., *Stripline Circuit Design*, Norwood, MA, Artech House, 1974.
Johnk, C.T.A., *Engineering Electromagnetic Fields and Waves*, John Wiley and Sons, 1973.
Kraus, J.D., *Electromagnetics*, McGraw-Hill, New York, 1984.
Kyocera Design Guidelines Multilayer Ceramic, CAT/1T8205TDN, Kyocera International, Inc., San Diego, CA.
Mohr, R.J., "Coupling Between Open Wires Over a Ground Plane," *IEEE Symp. EMC*, July 23–25, 1968.
Olsen, L.T., "Application of the Finite Element Method to Determine the Electrical Resistance, Inductance, Capacitance Parameters for the Circuit Package Environment," *Trans. IEEE*, Vol. CHMT-5, No. 4, December 1982.
Ott, H.W., *Noise Reduction Techniques in Electronic Systems*, New York, John Wiley and Sons, 1976.
Pantic, Z., and Mittra, R., "Quasi-TEM Analysis of Microwave Transmission Lines by the Finite-Element Method," *IEEE Trans.*, Vol. MTT-34, No. 11, November 1986.
Reference Data for Radio Engineers, Indianapolis, Howard W. Sams, 1968.
Rogers, W.E., *Introduction to Electric Fields*, New York, McGraw-Hill, 1950.
Schneider, M.V., "Microstrip Lines for Microwave Integrated Circuits," *Bell System Technical Journal*, Vol. 48, No. 5, May–June 1969.
Weber, E., *Electromagnetic Fields*, New York, John Wiley and Sons, 1954.
Zahn, M., *Electromagnetic Field Theory: A Problem Solving Approach*, Robert E. Krieger, Malabar, FL.

List of Symbols

Symbol	Definition
A	ampere
B	magnetic flux density, Wb/m²
C	capacitance, F
C_{ct}	crosstalk capacitance
C_m	mutual capacitance
C_{mct}	mutual crosstalk capacitance
°C	degrees centigrade (Celsius)
D	electric flux density, coulombs/m²
d	distance
dB	decibel
E	electric field intensity, V/m
E_{ia}	amplifier input voltage, V (for amplifier A)
E_{oa}	amplifier output voltage, V (for amplifier A)
ESR	equivalent series resistance
F	farad
G	conductance, S, Ω^{-1} (also called mhos, ℧)
G	gain, dimensionless
H	henry
H	magnetic field intensity, A/m
h	height
I	current, A
i	current, A
in	inch
J	current density, A/m²
K	fringing factor, dimensionless
K	crosstalk, dB
K_{C1}	capacitive fringing factor for microstrips, dimensionless
K_{C2}	capacitive fringing factor for striplines, dimensionless
K_{L1}	inductive fringing factor for microstrips, dimensionless
K_{L2}	inductive fringing factor for striplines, dimensionless
k	kilo (10^3)
L	inductance, H
L_m	mutual inductance, H
l	length

ln	logarithm, base $e = 2.71828\ldots$
log	logarithm, base 10
M	mega (10^6)
m	meter
m	milli (10^{-3})
N	number of turns
n	nano (10^{-9})
n_p	number of unit cells in parallel
n_s	number of unit cells in series
p	pico (10^{-12})
Q	resonance factor
Q	charge, coulomb
R	resistance, Ω
r, R	radius
RTN	common return for two or more power supply sources
S	siemen
t	thickness
U	magnetic potential, A
V	voltage, volts (V)
V	volt
V p-p	volts peak-to-peak, V
V rms	volts rms, V
Wb	weber
w	width
Z	impedance, Ω
Z_0	characteristic impedance, Ω
α	magnetic flux boundary ratio
β	electrostatic potential ratio
δ	depth of penetration, m
ε	dielectric constant, F/m (also called permittivity)
ε_0	dielectric constant in vacuum $= 10^{-9}/36\pi$ (F/m)
ε_r	relative dielectric constant, dimensionless
$\varepsilon_{r(\text{eff})}$	effective relative dielectric constant, dimensionless
Γ	geometrical factor, dimensionless
Λ	linear charge density, coulombs/m
μ	micro
μ	permeability, H/m
μ_0	permeability in vacuum $= 4\pi \times 10^{-7}$ H/m
μ_r	relative permeability, dimensionless
π	$3.14159\ldots$
p'	resistivity, Ω-cm
ρ	resistivity, Ω-m
σ	conductivity, $(\Omega\text{-m})^{-1}$
Φ	electrostatic potential, V
Ω	ohm
$\mho$	mhos (Ω^{-1})
ω	frequency, radian/s
Ψ	total electric flux, coulombs
Ψ_m	total magnetic flux, Wb
$\mathcal{R}$	reluctance, H^{-1}

Index

ac source, 171
Ancillary circuit elements, 165
Axis symmetry, 19

Capacitance
   between parallel, vertical, flat conductors, 48
   between circular conductors, 17, 32
      and a ground plane, 36
      with different radii, 34
   between a flat conductor and a ground plane, 55
   between a flat conductor and two ground planes, 71
   between horizontal, rectangular conductors of different widths, 51
   between long, parallel, horizontal, rectangular conductors, 51
   between two small spheres, 83
   of coaxial cables, 81
   crosstalk, C_{ct}, 125
      mutual, C_{mct}, 134, 138
   mutual, 114, 198
      four-conductor system, 66
      two-circular conductors near a ground plane, 39
      two-parallel, horizontal, flat conductors between two ground planes, 71
      two-parallel, horizontal, flat conductors near a ground plane, 62
   stray, 126
Capacitors
   ceramic, 155
   decoupling, 140
   electrolytic,
      aluminum, 153
      tantalum, 153
   mica, 155
   parallel plate, 15, 20, 27
   unit, 15
Cellular polyethylene, 82, 117
Ceramic module, 71, 73
Characteristic impedance versus land width,
   for buried conductors, 75
   for surface conductors, 124
Characteristic impedance, 22, 58, 74, 115, 121, 140
   long, circular conductors over a ground plane, 122
   of coaxial cables, 121
   long, flat, conductor between two ground planes, 123
   long, flat, conductors over a ground plane, 123
   long, parallel, circular conductors, 122
   long, parallel, horizontal, flat conductors, 123
   long, parallel, vertical, flat conductors, 123
   for various geometries, 121
Charge, 6
Coaxial cables, 81, 115
Coaxial cables, types, 82, 115
Conductance, 2, 23
Conductivity, 3, 23
Conductor perimeter, 54
Conductor resistance, 165
Crosstalk
   capacitive, 125
   capacitive coupling to summing junctions, 125, 133
      with ground plane, 125
   capacitive *versus* inductive, 135
   common ground coupling, 140
   due to power supplies, 149
   inductive, 135
Crosstalk analysis, Chapter 3, 125
Crosstalk factor,
   capacitive, K_c, 125, 133
   common ground, K_g, 140
   inductive, K_l, 135
Crosstalk factor, K, 125, 133
Current flux, density, 4
Current fringing, 5
Current regulating diode, 172
Current sinks, 172
Current sources, 172
Curvilinear squares, 17

dc source, 171
Decoupling capacitors, 140
Dielectric constant, 6, 24
   effective, 31, 53, 131, 189
Discrete components, 153
   capacitors, 153
   inductors, 158
Downstream power supply location, 140

Electric field, 2, 43
  intensity, 4, 8
  mapping, 15
Electric flux, density, 8
Electric potential, 5, 9
Elliptic integrals, 120
Elliptical, rectangular conductor, 20
Equipotential circles, 18
Equipotential surfaces, 9, 10, 15
Equivalent series resistance (ESR), 153
Experimental results, predicted *versus* measured, 177
Experiments and test data, 177
  capacitance and capacitive crosstalk, 178
    horizontal, flat conductors, EXP C-5A, 188
    horizontal, flat conductors with guard rings, EXP C-5B, 191
    horizontal, flat conductors with ground plane, EXP C-7, 198
    vertical, flat conductors, EXP C-4, 178
  ground return crosstalk, 206
    downstream power supplies, 209
    shared-ground, 206
    single-point ground, 208
  inductance and inductive crosstalk, 201
    four-conductor system mutual inductance, 201
  introduction, 177

Finite element analysis, 22, 80, 113
Flat cable, 32, 37, 85
Flux plotting, 15, 42
Flux tube boundaries, 19
  circular, 18
Flux tubes, 15
Formula Sets, Chapter, 31
Fringing factor, 185
Fringing factor, K, 21, 49, 56, 62, 63, 94, 110, 113, 131, 134
  versus ratio $2h/w$,
    for buried conductors, 71
    for surface conductors, 56
Fringing flux, 7, 48, 185
Gauss's law, 7
Geometrical factor, Γ, 22
Ground plane, 36, 39, 55, 62, 71, 79, 88, 89, 97, 100, 110, 113
Ground, shared, 206
Ground, single point, 143, 208
Guard lands, 191
Guard rings, 191

Impedance analyzer, 46

Inductance, 10, 23, 70
  between a circular conductor and a ground plane, 88
  between circular conductors, 85
  mutual, 39, 64, 133, 201
    four-conductor system, 101
    two conductors near a ground plane, 89
    two flat conductors between two ground planes, 113
    two flat conductors near a ground plane, 100
  self
    of a circular loop, 118
    coaxial cables, 115
    flat conductor between two ground planes, 110
    of a long flat conductor and a ground plane, 97
    long, parallel, vertical flat conductors, 92
    of a square loop, 118
    two parallel, flat conductors, 95
Inductor
  as a discrete component, 160
  parallel plate, 10, 28
Loop, voltage induced into, 90
$LRCZ_0$ analogy, 22, 92
Magnetic field intensity, 12
Magnetic flux, density, 12
Magnetic flux lines, 14, 42
Magnetic potential, 14

npn bipolar transistors, 173
Number of turns, 11

Operational amplifiers, 125, 178
Orthogonal lines, 15

Parallel plate structures, 26
Parallel resonance, 160
Permeability, 11, 24
pnp bipolar transistors, 172
Power supply characteristics
  downstream, 209
Printed wiring boards, 48
Pulsewidth modulator, 85
PVC insulation, 34, 38

Q-factor, 155

Relative dielectric constant, 21
Reluctance, 11, 13
Resistance
  equivalent output, 174
  ground plane, 169

Resistivity, 3, 166, 170
Resistors, 2
　parallel plate, 2, 27
Ripple voltage, 173

Scalars, 5, 9
Shared power supply bus, 140
Single-point ground analysis, 143
Solid polyethylene, 82, 117
Spikes, voltage, 175
Summing junction, 125
Switching type supplies, 174

Temperature coefficient for copper, 167
Test circuit boards, 178
Test equipment, 178
Two-dimensional geometry, 29

Unit cells, 15

Vectors, 5, 9, 13
Victim circuit, 128
Voltage potential, 4, 6
Voltage sources, 171

Zener diode, 172

THE AUTHOR

Charles S. Walker is currently a project staff engineer at Honeywell, Inc. He has served several roles at Honeywell since 1963, starting as Senior Development Engineer and becoming Project Supervisor in 1964 before undertaking his present role in 1982. He graduated from the University of Pennsylvania, Moore School of Electrical Engineering with an MSEE in 1962, and the University of Portland with a BSGE in 1952. He holds 15 patents.

Printed in the United States
59132LVS00002B